T0281159

Breaking Barriers

Breaking Barriers

Student Success in Community College Mathematics

Brian Cafarella

CRC Press
Taylor & Francis Group
Boca Raton London New York

CRC Press is an imprint of the
Taylor & Francis Group, an **informa** business

AN A K PETERS BOOK

First edition published 2021
by CRC Press
6000 Broken Sound Parkway NW, Suite 300, Boca Raton, FL 33487-2742

and by CRC Press
2 Park Square, Milton Park, Abingdon, Oxon, OX14 4RN

© 2021 Brian Cafarella

CRC Press is an imprint of Taylor & Francis Group, LLC

The right of Brian Cafarella to be identified as author of this work has
been asserted by him in accordance with sections 77 and 78 of the
Copyright, Designs and Patents Act 1988.

Reasonable efforts have been made to publish reliable data and
information, but the author and publisher cannot assume responsibility
for the validity of all materials or the consequences of their use. The
authors and publishers have attempted to trace the copyright holders
of all material reproduced in this publication and apologize to
copyright holders if permission to publish in this form has not been
obtained. If any copyright material has not been acknowledged please
write and let us know so we may rectify in any future reprint.

Except as permitted under U.S. Copyright Law, no part of this book
may be reprinted, reproduced, transmitted, or utilized in any form by
any electronic, mechanical, or other means, now known or hereafter
invented, including photocopying, microfilming, and recording, or in
any information storage or retrieval system, without written
permission from the publishers.

For permission to photocopy or use material electronically from this
work, access www.copyright.com or contact the Copyright Clearance
Center, Inc. (CCC), 222 Rosewood Drive, Danvers, MA 01923, 978-750-
8400. For works that are not available on CCC please contact
mpkbookspermissions@tandf.co.uk

Trademark notice: Product or corporate names may be trademarks or
registered trademarks and are used only for identification and
explanation without intent to infringe.

ISBN: 978-1-032-02425-7 (hbk)
ISBN: 978-1-032-00797-7 (pbk)
ISBN: 978-1-003-17580-3 (ebk)

Typeset in Palatino
by MPS Limited, Dehradun

This is for my beautiful wife Lisa and my wonderful son Gavin. You bring so much joy to my life every day.

This is for my parents, Margaret and John, who always believed in me and never gave up on me.

This is for my in-laws Ethel and Roger, who do so much for our family.

This is for my two math professors during my freshman year of college, Dr. Constance Beck and Mrs. Toby Grossfield. I entered community college very unsure of my own abilities. You both believed in me and taught me so much. To this day, I still employ teaching techniques that I learned from both of you.

Contents

Preface

As a community college professor, I see that students struggle in mathematics. More specifically, I see a great deal of anxiety, dislike, and overall disinterest from students regarding mathematics. Throughout my initial years of teaching, I recall attending many crisis meetings that focused on the lack of student success in mathematics. The tone of each meeting was the same. We were given quantitative data that displayed abysmal student success rates. There was discussion regarding how to improve student success in math, and oftentimes we were presented with various initiatives to implement into our classes. Despite this, students continued to struggle. I became frustrated over the continuous cycle of getting beaten up with statistics while students continued to have trouble in mathematics. More specifically, I wanted to understand more about the complexity surrounding the issue of student difficulties in community college math.

Like other math faculty, I have been involved in various initiatives to improve student success in math. This can be intimidating. When implementing a new practice in my classroom, I want to be sure that I am helping and not hindering my students' success. However, I understand the need for change. As time progresses, we must evolve and alter some of our teaching practices to help students. While I do not fear change, I do fear chaos. Chaos can erupt when an initiative or teaching practice is improperly implemented or is simply a poor fit. This creates anxiety and frustration for students, and consequently they do not learn. Therefore, it is imperative that we holistically understand pedagogical initiatives and practices prior to implementation.

While completing my doctorate, I became interested in research, particularly qualitative research. Quantitative research, which typically provides descriptive data or statistical outcomes based on an observational or experimental study, is imperative to measure whether an academic discipline, such as community college mathematics, is effective or is in good standing. Qualitative research, however, takes an inductive approach to studying a phenomenon. More specifically, qualitative research focuses on the "how" and the "why". To improve the student experience in community college math, I began to conduct qualitative research on various teaching methods that community college math faculty utilized with students. My research also focused on the faculty experiences and perspectives regarding such methods and practices. However, I recently had a realization that my understanding of how to improve student success in community college math should also focus on student experiences.

In this book, I will present a study that focused on students who struggled in math. These are twenty-five students who reported struggling with math at various points in their lives; most, in fact, struggled with math their entire lives.

These students entered community college with a great deal of disgust and anxiety toward math. However, all twenty-five students eventually succeeded in fulfilling their college-level math requirements. What were their math experiences prior to entering community college? What led to both success and failure in their math courses? More specifically, what were the common themes that led to success and failure that emerged from students' experiences? In summation, I wanted to gain and present a better understanding of the community college student who struggles in math and how we can break the community college math barriers.

Acknowledgment

I would like to thank the participants of this study for the time you took out of your schedules to complete the interviews. Your commitment to this project will help other students succeed in community college mathematics.

I would like to show my appreciation to the math faculty and administration at Sinclair Community College. I could not ask for a better group of co-workers. Your dedication to our students is contagious, and you make me constantly strive to be a better teacher.

I wish to thank Dr. Darla Twale for patiently answering my many questions and providing me with great advice throughout this entire process.

I wish to thank CRC Press for working with me and believing in this project.

Author

Brian Cafarella, PhD, is a mathematics professor at Sinclair Community College in Dayton, Ohio. He has taught a variety of courses ranging from developmental math to pre-calculus. Dr. Cafarella is a past recipient of the Roueche Award for teaching excellence. He is also a past recipient of the Ohio Magazine Award for excellence in education.

Dr. Cafarella has published his articles in several peer-reviewed journals. His articles have focused on implementing best practices in developmental math and various math pathways for community college students. Additionally, Dr. Cafarella was the recipient of the Article of the Year Award for his article, "Acceleration and Compression in Developmental Mathematics: Faculty Viewpoints" in the *Journal of Developmental Education*.

1

Math Is a Four-Letter Word

When new acquaintances ask what I do for a living, I respond that I teach at a community college. This generally produces increasingly interested facial expressions, and they ask what I teach. As soon I respond with the four-letter word of math, their eyes generally roll back as their heads drop, and sometimes a look of fear crosses their faces. As a community college math professor, I also see these reactions from students in my classroom.

I must remind myself that while entering community college as a student, math was my least favorite subject. I did not have a fear of math; I simply had not done well in math in high school. I had no interest in the subject whatsoever, and I was confident that I would fail math. I felt lost in many math classes for many years. Why would college be any different? Like many other students, I placed into basic algebra, which is below college level.

For me, high school was a bad experience. I was unhappy for many reasons; ultimately, I did not fit in, and I did not take my schoolwork seriously. Consequently, I slid through high school with a C average. My junior year ended, and for many students, this was the beginning of the college application process, which was grounded in years of hard work and scholastic achievements. For myself, I was simply relieved to be finished with another dreadful year of school, and I was looking forward to graduating in one year.

A few days after I completed my junior year, I was at a local Walmart in the candy aisle. As I was deciding whether to purchase jelly beans or mints, I noticed an elderly man to my right. He was an employee and was stacking and sorting the bags of candy. For the first time in my life, I was able to see beyond the present moment and into my future. I realized that high school would end in one year , I would be free; however, I would be on my own to face the world as an irresponsible high school graduate with a mediocre grade point average. Was this my future? Was I destined to continue to live in the same town working at Walmart for minimum wage? At once, I decided that I needed a better plan or at least some plan.

Starting at a university after high school was not an option. I did not possess the grades, and quite frankly, I knew that I was not ready to live away at a university, so I chose a community college. Truthfully, I did not expect to make it through a full year. It had been a long time since I was a good student. As luck would have it, math was my first class on my first day. Again, I was not afraid of math; I just did not foresee success. As even more luck would have it, I was late to my first class on my first day as I had not anticipated the lack of available parking spaces. I ran into class in a panic and sat down several minutes late. My teacher looked at me, smiled, and said good morning as she handed me a syllabus. My basic algebra teacher, Dr. Connie, was excellent. She explained math unlike anyone else had. That, combined with my commitment to take school seriously, led to my excelling in the course.

After completing my college-level math course, I decided to take a part-time job as a student-tutor the lower level math classes at my community college. I fell in love with teaching, and I valued working with community college students who struggled in math, and I showed them different ways to conquer a problem. I knew what I wanted to do with the rest of my life: to become a community college math professor.

I continued taking math classes and eventually earned my bachelor's and master's degrees, all the while focused on my goal to teach full time at a community college. I achieved a tenure-track position at my current community college, and I thought it would be relatively simple. My students would enter, having had difficult experiences in math and school, but like me, they decided to take school seriously and enter community college. I would teach them well, and they would succeed and possibly even develop an appreciation for math. This seemed both logical and realistic.

While I have enjoyed—and still enjoy—teaching math and love working with my students, it has not been that simple. Students struggle mightily in both developmental and introd][uctory college-level math. More specifically, through the years, I have seen students appearing unprepared and unmotivated, even taking the same math course multiple times. As a professional who devotes much time to my teaching craft, this has always bothered me.

The Endeavor to Fail

Over the years, I realized that students' struggles in community college math are a national issue. Wang et al. (2017) reported that only 45.59% of students who try a college-level math class are successful. This is certainly problematic; however, the larger issue is that nearly 60% of students who enroll in a community college place into developmental mathematics (U.S. Department of Education, 2017). Boylan and Bonham (2007) referred to developmental courses as those below college level. Wang et al. (2017) stated that for students who test into developmental math, only 31% complete their developmental math course sequence. When the developmental course sequence consists of at least three courses, the success rate lowers to 22%.

Unfortunately, poor statistics have been linked to developmental math since the 1990s. Bahr (2008) reported that 81.5% of students who took a developmental mathematics course did not complete a community college degree or transfer to another school. In 2006, based on a national study, Attewell et al. stated that only 30% of students who enroll in developmental math courses are successful. In 1997, Boylan reported that less than 10% of the students who are unsuccessful in their developmental coursework stay in school.

Lack of success in developmental math can negatively affect students in multiple ways. It grows costly for students to repeat courses, and they may accumulate debt through this process. Students receiving financial aid are generally required to carry a minimum number of credit hours and uphold a certain grade point average. Therefore, when students repeatedly fail their mathematics courses, the government terminates their financial aid (Bailey, 2009). Students also begin to doubt their academic ability, and this may cause them to withdraw entirely from college (Boylan, 2011).

Community Colleges

I interviewed students who completed their math requirements at various community colleges. I provided a brief description of each community college in chapter 3. However, below is a brief background on the American community college.

While two-year colleges have existed in America since the 1880s, public access community colleges, as they are known today, grew in the 1960s. In fact, Geiger (2005) conveyed that between 1965 and 1972, community colleges opened at a rate of one per week. The surge of community colleges stemmed from the Higher Education Act of 1965 and Lyndon Johnson's War on Poverty (Gladieux et al., 2005). More specifically, American education began to focus on the poor and underserved (Boylan, 1988). This was higher education's shift to serving the universal population.

Community colleges generally consist of diverse student populations. Community colleges tend to offer something for most students, as they offer a wide variety of associate degrees as well as certificates. Most community colleges do not require a high school diploma or even an equivalency diploma.

Placement in Community College Math

Most community colleges require incoming students to complete a standardized placement exam, and the exam results indicate the level of mathematics at which a student must begin. Institutions choose their own placement tool; however, ACCUPLACER is commonly utilized to place students. Reports regarding predictive validity, a student's subsequent performance in math classes, have been mixed for ACCUPLACER. James (2006) found strong predictive validity for mathematics classes. However, Medhanie et al. (2012) reported weak predictive validity. Community colleges may also employ multiple measures to place students. More specifically, in addition to or instead of a placement exam, colleges may rely on a student's ACT or SAT score to determine placement. Multiple measures may also rely on a student's GPA or even prior math background. The use of multiple measures has proven to be effective as placement scores. Ngo and Kwon (2015) found that students who placed into math courses using multiple measures performed as well as their peers who were placed into math classes using traditional placement scores.

Does a student's performance in high school math impact their placement in community college math? The answer varies. Zelkowski (2010) reported that fewer years of math in high school result in lower math placement in community college. However, in their study, Benken et al. (2015) found that two-thirds of the students who required remedial math completed four years of high school. Furthermore, over 20% of these students completed higher level math courses such as pre-calculus or even calculus.

Developmental Math, College-Level Math, and the Courses in Between

I will be referring to developmental math throughout this book as it was an imperative part of the study. The content in developmental math courses is the content employed in secondary or even elementary school. Such content includes pre-algebra, elementary algebra, and intermediate algebra. Developmental math also includes courses with arithmetic content. However, the U.S. Department of Education has become stringent about providing financial aid for classes below the ninth grade level. Therefore, many community colleges may not offer stand-alone arithmetic courses. Students generally do not receive academic, or graduation, credit for completing these courses, as it is unlikely that any developmental math course will satisfy a two-year or especially a four-year degree.

Introductory college-level math courses not only satisfy a two-year degree, but some satisfy the math requirements of a four-year school for those who transfer. Such courses include college algebra, quantitative reasoning, introduction to statistics, and teacher preparatory courses. I describe the content of these courses in Chapter 8.

Many community colleges offer math courses for allied health majors and introductory business math courses. Although these courses may satisfy two-year degrees in allied health and business, respectively, they generally do not transfer to four-year schools. The content in these courses consists of arithmetic word problem applications, which is not strong enough to be considered college-level math. However, since the difficulty level for math for allied health majors and introductory business math falls somewhere between developmental math and introductory college-level math, the common themes for success and failure that I present could be extrapolated to these courses as well.

Looking Ahead

Clearly, student success in community college math, or the lack thereof, has been a salient issue for some time. Therefore, I set out to gain an in-depth understanding of students' experiences in community college math, especially those who have struggled. However, I had some preliminary questions. Why do other educators and researchers believe so many students struggle in community college math? What have colleges done to address this issue? Since developmental math is attached to so many negative statistics, do researchers and educators feel this discipline is effective? I will address these questions in the next chapter.

References

Attewell, P., Lavin, D., Domina, T., & Levey, T. (2006). New evidence on college remediation. *Journal of Developmental Education, 77*(5), 886–924.

Bahr, P. R. (2008). Does remediation work? A comparative analysis of academic attainment among community college students. *Research in Higher Education, 49*(5), 420–450. https://doi:10.1007/s11162-008-9089-4

Bailey, T. (2009). Challenge and opportunity: Rethinking the role of developmental education in community college. *New Directions for Community Colleges, 45*, 11–30.

Benken, B. M., Ramirez, J., Li, X., & Wetendorf, S. (2015). Developmental mathematics success: Impact of students' knowledge and attitudes. *Journal of Developmental Education, 38*(2), 14–22. http://www.jstor.org/stable/24614042

Boylan, H. R. (1988). The historical roots of developmental education part III. *Research in Developmental Education, 5*(3). 1–14.

Boylan, H. R. (2011). Improving success in developmental mathematics: An interview with Paul Nolting. *Journal of Developmental Education, 34*(3), 12–41.

Boylan, H. R. & Bonham, B. S. (2007). 30 years of developmental education: A retrospective. *Journal of Developmental Education, 30*(3), 2–4.

Geiger, R. L. (2005). The ten generations of American higher education. In P. G. Altbach, R. O. Berdahl, & P. J. Gumport (Eds.), *American higher education in the 220 twenty-first century: Social, political, and economic challenges* (pp. 36–70). Baltimore: The Johns Hopkins University Press.

Gladieux, L. E., King, J. E., & Corrigan, M. E. (2005). The federal government and higher education. In P. G. Altbach, R. O. Berdahl, & P. J. Gumport (Eds.), *American higher education in the twenty-first century: Social, political, and economic challenges* (pp. 163–197). Baltimore: The Johns Hopkins University Press.

James, C. (2006). ACCUPLACER online: Accurate placement tools for developmental programs? *Journal of Developmental Education, 30*(2), 2–8.

Medhanie, A. G., Dupuis, D. N., LeBeau, B. C., Harwell, M. R. (2012). The role of the ACCUPLACER mathematics placement test on a student's first college mathematics course. *Educational and Psychological Measurement, 72*(2), 332–351. https://doi.org/10.1177/0013164411417620

Ngo, F., Kwon, W. (2015). Using multiple measures to make math placement decisions: Implications for access and success in community colleges. *Research in Higher Education, 56*(5), 442–470. http://dx.doi.org/10.1007/s11162-014-9352-9

U.S. Department of Education, Office of Planning, Evaluation and Policy Development. (2017). *Developmental Education: Challenges and Strategies for Reform.* Washington, DC: U.S. Department of Education.

Wang, X., Wang, Y., Wickersham, K., Sun, N., Chan, H. (2017). Math requirement fulfillment and educational success of community college students: A matter of when. *Community College Review, 45*(2), 99–118. https://doi.org/10.1177/0091552116682829

Zelkowski, J. (2010). Secondary mathematics: Four credits, bock schedules, continuous enrollments? What maximizes college readiness? *The Mathematics Educator, 20*, 8–21. https://www.jstor.org/stable/23100400

2

The Framework for Developmental and Introductory College-Level Math: Why Are So Many Students Unsuccessful?

Attempting to identify why so many community college students struggle in their endeavor to complete their math requirements is not a simple task. Research has shown a variety of reasons.

Lack of Prerequisite Skills

Mathematics, by default, is a progressive and linear discipline (Boylan, 2011). However, community college students often attempt math courses in which their prerequisite skills are deficient (Boylan, 2011; Dahlke, 2011). Dahlke (2011) asserted that passing a prior math class does not indicate students possess the necessary foundation for the current class. More specifically, a student may achieve a low C grade, which would not provide enough knowledge base for higher level math. Dahlke further postulated that a time lapse can be an issue as well. If students wait too long after completing a math course to take a subsequent math class, they may become rusty or even forget the required skills.

Others argue that the prerequisite issue runs deeper. Many students enter college lacking basic arithmetic skills (Stigler et al., 2010; Xu & Dadgar, 2017). This may place students in lower level developmental math, and even if students place in a higher level math class, these basic mathematical gaps thwart students from acquiring sophisticated material. Despite numerous calculator debates over time, some educators assert that students become calculator dependent as early as elementary school (Boylan, 2011). Furthermore, the dependency on the calculator inhibits logical thinking skills, which are imperative for success in math (Winerip, 2011).

Math Anxiety

Many students develop an intense fear of mathematics, otherwise known as math anxiety. Boylan (2011) found that this anxiety becomes so debilitating for some students that they cannot concentrate in class, complete math assignments, or take exams. Diaz (2010) suggested this anxiety may stem from classroom experiences such as poor performance in math or intimidating teachers. Jameson and Fusco (2013) reported that this is especially the case for returning or nontraditional students. Nontraditional students report lower self-efficacy in mathematics, which can hinder their performance. Boylan (2011) added that advances in technology in education can add to the anxiety for nontraditional students. These students have enough anxiety as it is, and learning math through technology can make the subject even more intimidating.

Affective Behaviors

Over time, faculty and others have postulated that various student behaviors have thwarted their road to success in college-level math. Smith et al. (1996) and Merseth (2011) have identified poor attendance as a key factor. Considering the linear progression of mathematics, poor attendance leads to gaps in the knowledge base, which in turn lead to failure. Poor attendance has been attributed to apathy. Students do not see the significance of the content in their math classes or how such content relates to their everyday lives (Wheland et al., 2003). However, poor attendance has also been ascribed to outside factors such as work or family obligations taking priority over education (Immerwahr et al., 2005). This can especially be the case for students who must complete long developmental math sequences, which can become costly and time consuming (Xu & Dadgar, 2017).

Overall, poor work habits have been blamed for lack of success in community college math. Other behaviors include failure to take notes in class and poor study habits (Pang, 2010; Smith et al., 1996; Stigler et al., 2010). Howard and Whitaker (2011) found that students' dislike and fear of math led them to avoid the subject. Stigler et al. added that developmental math students often memorize formulas and facts, but they fail to make the deeper connections needed to succeed in college-level math.

It is noteworthy, however, that students who are unsuccessful in their developmental math sequence often exhibit low academic performance in their coursework in general (Bahr, 2013). This may indicate that these students display these affective behaviors in other courses as well.

The Pathway to Negative External Attention

Community colleges are public institutions of higher education that rely heavily on state funding to operate (Weisbrod et al., 2008). Unlike four-year state schools, community colleges offer low tuition (Phillippe & Sullivan, 2005), which makes state funding salient. However, since the early 2000s, state legislatures have become increasingly frustrated with community college math programs. Considering the large number of students who place into developmental math and never complete their developmental math requirements, this discipline has come under intense scrutiny.

Where Did Developmental Math Come From?

I was attending a state conference that focused on poor retention rates in math. At one break out session, a frustrated administrator, from a community college asked, "Where did developmental math even come from? I just put my own kids in college math after high school. There was no developmental math!" The reality is that developmental math was in existence when this gentleman's children attended college, and developmental education, in general, long predates the existence of the community college.

The practice of developmental education, which serves the underprepared student mostly, can be traced back to the inception of higher education. This is when Harvard College opened in 1636 (Boylan & White, 1987). However, at that time, most college courses were instructed in Latin, and most books were written primarily in Latin (Thelin, 2004). However, most colonists were not fluent in Latin and were consequently underprepared for college (Boylan & White, 1987). Harvard College assigned young men, labeled "tutors", who already possessed baccalaureate degrees and were preparing for careers in the ministry, to assist the underprepared students alongside the tutors' regular coursework (Brubacher & Rudy 1976). Accordingly, programs for underprepared college students were labeled "tutoring" (Arendale, 2002).

After the American Revolution, courses in higher education were instructed in English; however, students were still underprepared and needed extra assistance (Brubacher & Rudy 1976). The tutors were not able to keep pace with this demand (Boylan & White, 1987). Consequently, in 1849, the University of Wisconsin established a program with stand-alone courses in reading, writing, and math for underprepared students. This was the nation's first formal remedial program (Brier, 1986). At this point, such courses were referred to as precollegiate, college preparatory, or remedial (Arendale, 2002). Other colleges soon followed the University of Wisconsin's model for remedial education.

In 1890, to reduce the need for remediation and raise academic standards, the College Entrance Examination Board was established. Up to that point, there were no academic requirements to enter college. Higher education was strictly tuition driven, and anyone who had enough money could attend college (Brubacher & Rudy, 1976). However, students still needed remediation. As two-year junior colleges emerged, these institutions began serving underprepared students as well (Boylan, 1987).

The need for remediation continued throughout the twentieth century. This intensified in 1944 when the US government passed the Servicemen's Readjustment Act, also known as the GI Bill of Rights. This was due to the government's concern regarding the millions of veterans who were returning from World War II to potential unemployment. Therefore, the GI Bill allocated millions of dollars for the education and training of these war veterans (Thelin, 2004). Consequently, several millions of returning American war veterans used the GI Bill to register in college (Olson, 1974). Many of the students who attended college via the GI Bill were unprepared and required extra assistance.

The need for remedial education further increased with the growth of open access community colleges in the 1960s and early 1970s. In fact, the name for these courses shifted from "remedial" to "developmental". In line with serving the underprepared and underrepresented student, the term referred to focusing on student development in addition to preparing the student for college-level content (Arendale, 2002; Boylan & Bonham, 2007). Throughout the 1970s and 1980s, developmental education became recognized as an academic discipline in higher education. In 1976, the National Association of Developmental Education (NADE) was financed by the W.K. Kellogg Foundation at Appalachian State University. In 1980, NADE funded the formation of the Kellogg Institute for the Training and Certification of Developmental Educators at Appalachian State University. This was a formal program for developmental educators to study methods for assisting underprepared students (Spann, 1996).

Developmental math was a major part of community colleges. In 1975, Baldwin reported that 91% of all two-year colleges offered developmental math. However, little is known regarding student success rates. Although developmental math courses were offered, little was done to evaluate the effectiveness of such courses (Rutschow & Schneider, 2011). However, this began to change toward the end of the twentieth century.

The States Take Notice

During the inaugural period of community colleges, states funded these institutions almost solely on student full-time equivalent (FTE). FTE is calculated by the sum of all credits carried by students divided by the number of credits in a full-time load. More specifically, the higher an institution's enrollment, the more state funding the institution receives. Again, with a focus on serving the underrepresented students, states generously supported community colleges (Vaughan, 1985). Toward the end of the twentieth century, however, this began to change. In fact, the American Council on Education (2004) conveyed that between 1980 and 2000 the state support for public four-year colleges and community colleges decreased steadily. Other expenses, namely Medicaid and prison costs, started to pull at state budgets. Rizzo (2006) asserted, "Medicaid costs have skyrocketed as a result of large increases in caseloads, escalating prescription costs, and lagging support from the federal government" (p. 5). Additionally, the American Council on Education (2004) reported that state spending on corrections increased by six times the rate of spending on higher education between 1985 and 2000. Consequently, state legislatures began examining student success rates at community colleges to justify funding.

Since the early 2000s, both state and federal legislatures have become increasingly frustrated with community college math success rates. Moreover, legislatures began to pressure these institutions and made them more accountable for student success and completion rates (Arendale, 2003; Boylan, 2008). More specifically, states began to impose funding formulas on community colleges. Such formulas placed more emphasis on student completion rates as opposed to FTEs (Li et al., 2018).

The cost of developmental education came into the spotlight as well. In 2010, the state higher education executive officers reported that the cost for developmental education was as high as $3 billion. Although this number seems high, the state higher education executive officers also conveyed that government spending on higher education was approximately $140 billion. Consequently, Goudas and Boylan (2012) suggested that the cost of developmental education is relatively low, as it accounts for less than 2% of the higher education budget. Nonetheless, this fueled changes in the community college math paradigm.

Addressing the Math Issue

Since the start of the twenty-first century, community colleges have developed various initiatives to increase student completion rates in mathematics. In 2003, the Lumina Foundation, a private, Indianapolis-based group, initiated discussions with various national educational experts focusing on community college education. The Lumina Foundation provided several millions of dollars to community colleges across the country to fund Achieving the Dream: Community Colleges Count (Ashburn, 2007). Immerwahr et al. (2005) posited that the program's primary focus was on improving the "achievement of community college students, especially those facing the greatest obstacles" (p. 2). Colleges that participated in Achieving the Dream examined their success rates and overall data and developed initiatives to improve student success. A focus of Achieving the Dream is improving success in gatekeeper courses, which certainly includes mathematics.

As the aughts (the first decade of the twenty-first century) ended, the initiatives to improve success rates in community college math became more uniform. Community colleges began to search for ways to compress or accelerate the developmental math course sequence so that students could reach their college-level math course in less time. The Bill and Melinda Gates Foundation invested several millions of dollars for schools to develop groundbreaking ways to accelerate students through their developmental math coursework (Ashburn, 2007). The abysmal success rates in developmental math courses were a driving factor. Data continued to emerge that supported acceleration. Bailey et al. (2010), for example, reported that only 33% of students who place into developmental math courses complete their required work within three years. Additionally, only 17% of students successfully complete a developmental math sequence of at least three courses. Also, students who place into the lowest level developmental math class have the least chance of completing the sequence (Bahr, 2013; Xu & Dadgar, 2017).

Throughout the 2010s, some states began scaling back stand-alone developmental math courses. This was likely due to pressure from Complete College America (CCA), a nonprofit advocacy group that the Gates Foundation funded. CCA has urged state lawmakers to reduce or eliminate remedial courses so that students can progress into their college-level courses at a quicker rate (Mangan, 2013). In 2014, many Florida community colleges allowed students to start in a college-level math course, and enrollment in developmental math was optional. Other states, such as Connecticut, Tennessee, and California, have followed suit.

Is Developmental Math Effective?

Developmental math has received a great deal of criticism owing to the length of course sequence, the cost concerns, and low success rates. This raises the question: Do developmental math courses prepare students for college-level math? The answer is complex. Quarles and Davis (2017) found that developmental math classes do not sufficiently prepare students for college-level math, as these courses teach more procedures but do not give students a deeper conceptual understanding needed for college-level math. However, Wheeler and Bray (2017) found that developmental math classes help improve the odds of graduating, as students who prospered in developmental math were successful in college-level math. Also, when developmental math was optional, Park et al. (2017) reported that students who opted for developmental support were more successful in their math classes.

Initiatives to Improve Success in Math

As states have pressured community colleges to improve success rates in math and create more seamless pathways for college completion, there have been various initiatives focusing on math improvement.

Emporium Model

The most common practice of acceleration is the use of an emporium-style classroom. The emporium model allows students to work at their own pace using software programs such as MyLabMath, ALEKS, or HAWKES to complete their developmental math coursework. This modality contrasts with lecture-based instruction, which had been the most common form of instruction for community college mathematics (Boylan, 2002), as there is little to no lecture from the instructor. Instead, math videos help students understand the content. The lead instructor serves more as a facilitator and, along with in-class tutors, works with students individually. The emporium model of instruction originated in the late 1990s at Virginia Polytechnical and State University. Throughout the aughts, roughly 120 institutions adopted this model, several as part of the Achieving the Dream initiative (Twigg, 2011).

The emporium model addressed the issues of cost and sequence longevity regarding community college math. Since emporium classes are run in computer labs with a larger class size, faculty costs are reduced. Additionally, students can accelerate through their math courses, using software programs by focusing solely on these deficiencies and complete their developmental math coursework at a quicker pace.

The overall findings from the emporium model have been mixed. The National Center for Academic Transition has posited that implementing the emporium model improved success rates in developmental math courses (Twigg, 2011). However, in their study, Childers and Lu (2017) reported only 33% success rates for students who attempted the emporium model. Kozakowski (2019) also reported that students who endeavored the emporium model were less successful than those in traditional instruction and were less likely to persist in college.

Twigg (2011) elaborated on some of the beneficial characteristics of the emporium model. Such characteristics include more time on task in that students spend class time working on math as opposed to listening to a lecture; students can focus completely on their math insufficiencies; and students are more likely to get assistance in a self-paced computer lab as opposed to a traditional classroom. Beamer (2020), however, reported that while students receive individual assistance, instructors often report feeling overwhelmed and are not able to give each student the required attention.

A concern regarding the emporium model is whether this modality sufficiently prepares students for college-level mathematics. Childers and Lu (2017) asserted that only 58% of the students who passed developmental math via the emporium model were successful in college-level mathematics. Ariovich and Walker (2014) found that computer-based instruction does not provide a deep enough conceptual foundation for college-level math readiness. Beamer (2020) elaborated that computer-based instruction allows students to employ strategies such as guessing and multiple attempts. Such strategies, however, may hinder students' foundation for college-level math.

Alternate Math Pathways

Historically, college algebra has been the primary course that students must pass to earn credit that both satisfies a degree and is deemed transferrable to a four-year institution. As the search for methods to improve mathematics progressed, college algebra came under scrutiny. Cohen and Kelly (2020) reported that 68% of students who pass their developmental math courses are still unsuccessful in college algebra. Gordon (2008) asserted that college algebra served students who were on the road to calculus; however, this course did not benefit the non-STEM (Science Technology Engineering Mathematics) majors. Boylan (2011) elaborated by questioning how an algebra topic such as dividing polynomials would serve a nursing student. Consequently, change agents began searching for alternate pathways for the non-STEM major.

In 2009, the Carnegie Foundation for the Advancement of Teaching started a $13 million initiative to increase student success rates in math. During the 2010s, this led to two alternative pathways for non-STEM majors: Statway and Quantway (Merseth, 2011). The Carnegie Foundation also inaugurated a networked improvement community (NIC), which is a national group of mathematical academic representatives, community college administrators and faculty, educational researchers, and change agents (Bryk et al., 2015). This organization worked to develop content, effective pedagogical practices, and overall learning outcomes for Statway and Quantway.

Statway

Huang and Yamada (2017) described Statway as an alternative to the traditional developmental math and college algebra sequence where students can achieve college credit within one year. Statway consists of a developmental component, which specifically prepares students for the content of a traditional college-level statistics course. Yamada and Bryk (2016) emphasized that Statway utilizes problem-solving techniques and various methods of student engagement in the developmental component. Initially, students would take the developmental component the first semester and the college-level course in the second semester.

Yamada and Bryk (2016) reported positive findings during the first few years of Statway implementation. Students are earning college credit at a quicker pace, and fewer students are getting lost in the developmental course sequence. Huang and Yamada added that Statway provides more relevance to real-world applications compared to the traditional developmental and algebra sequence.

Quantway

Like Statway, Quantway began as a one-year alternative pathway for the non-STEM students. Quantway 1 serves as the developmental part of the Quantway sequence where students focus on basic prerequisite skills for the college-level portion, such as numerical skills, proportional reasoning, and algebraic reasoning. Quantway 2, which is generally labeled quantitative reasoning (QR), is the college-level portion of Quantway. Elrod (2014, para 7) defined QR as "the application of basic mathematics skills, such as algebra, to the analysis and interpretation of real-world quantitative information in the context of a discipline or an interdisciplinary problem to draw conclusions that are relevant to students in their daily lives". Howington et al. (2015) added that the QR curriculum "gets students through their math requirements in a shorter time while providing them the necessary mathematical and QR skills both to complete their college degrees and to use in their daily lives" (p. 2). Students focus on topics such as numeracy, modeling, and statistics. More specifically, students gain experience in real-world applications such as computing debt-to-income ratios, depreciation, and simple and compound interest.

Quantway positively influenced higher education during the 2010s. Based on studies conducted from 2011 through 2017 that compared the Quantway pathway to the traditional developmental math sequence, student success rates in Quantway 1 were three to four times better than rates for the traditional developmental math sequence, and students reached a college-level course in half the time (Huang, 2018). This concurred with the findings of Yamada et al. (2018) that Quantway 1 positively influenced success in fulfilling developmental math requirements. Huang also asserted that out of 814 Quantway students who studied during the 2016–2017 academic year, 77% were successful and earned college-level math credit.

Educational organizations, the Carnegie Foundation and the Charles A. Dana Center, were heavily involved in the development of Quantway. Both organizations have stressed two concepts in this course: the development of critical thinking skills for students in this course, and that the primary pedagogical practice for QR classes should be inquiry and group-based, as opposed to traditional instructor-led lecture. Elrod (2014) emphasized that the development of critical thinking skills is imperative to undergraduate education and for job skills. Group-based instruction encourages critical thinking by allowing students to work collaboratively and struggle through problems. Group-based instruction has had a constructive influence on student learning. More specifically, group-based instruction has led to improved communication among students as well as giving students a sense of belonging and community (Altose, 2018; Clyburn, 2013). Cafarella (2020) reported that instructors who teach the QR course are often pressured to employ group-based instruction as the primary learning modality, which can be disadvantageous to come students, as this type of instruction is not an ideal fit for every student. Cafarella recommended that QR faculty should have the freedom to choose the pedagogical approach that is suitable for the class.

Corequisite Model

The Statway and Quantway pathways shortened some students' endeavors to college-level math. The pathways vary between institutions, but students can reach college-level statistics or QR in a shorter time compared to college algebra. More specifically, most community colleges do not require the full developmental math sequence to reach either course. However, community colleges continue searching for methods to further shorten math pathways. Over time, the Community College of Baltimore developed the Accelerated Learning Program where students could enroll in college-level courses while completing their developmental requirements (Adams, 2017). This became known as corequisite education, and during the 2010s, various institutions began implementing corequisite pathways for Quantway and Statway. Students would enroll in the preparatory course, sometimes known as a booster course, concurrently attempting to complete college-level introductory statistics or QR.

Carnegie (2020) conveyed positive early results for both the Statway and Quantway corequisite pathways. More specifically, in a study consisting of 410 students across six community colleges, student success rates in either Statway or Quantway more than tripled the success rates in a developmental math sequence. Students also earned college-level math credit in less than a quarter of the time.

Individual institutions set the prerequisites and placement scores required for Quantway and Statway. However, Howington et al. (2015) asserted that these pathways target students who place at least two levels lower than a college-level math course. Cafarella (2020) reported that setting a proper prerequisite for corequisite classes was imperative to ensure students possessed an adequate incoming skill level. Default placement for Quantway and Statway corequisites, where there was no prerequisite, led to student struggles, as these students lacked basic math skills.

Distance Learning

The fully online format for developmental and introductory college-level math has steadily grown because of the rise of the online learning format, in general, during the twenty-first century. Between 1999 and 2005, the number of students taking at least one online class soared from 744,000 to over three million (Weisbrod et al., 2008). In 2018, the Online Learning Consortium reported that there has been a steady increase in online enrollment, and nearly one-third of higher education students take at least one online course.

In limited research, students perform lower in the online format for developmental and introductory college-level courses as opposed to the face-to-face format (Ashby et al., 2011; Francis et al., 2019). While online courses offer convenience, this format can be challenging, particularly to developmental math students. For example, developmental mathematics students lack the time management skills and independent learning capability that are salient in the distance learning format (Phillip, 2011). Wadsworth et al. (2007) postulated that students need precise learning tactics, self-efficacy, concentration, information processing, and self-testing skills to flourish in an online developmental mathematics course.

Other Practices

Throughout the aughts, learning communities became popular in higher education and in community college mathematics. Tinto (1998) defined learning communities as "a kind or co-registration or block scheduling that enables students to take courses together. The same students register for two or more courses, forming a sort of study team" (p. 169). Since two courses are linked together, students explore topics from both courses and attempt to identify how the courses relate. Tinto posited that students who make connections with their peers are more likely to persist in college. In a learning community, traditional lecture is supplemented with collaborative learning and group discussion.

During the 2010s, the use of learning communities decreased. Some institutions reported that higher success rates in math class were part of a learning community as well as students feeling more academically and personally supported (Visher et al., 2010; Weissman et al., 2011). However, Ashburn (2007) asserted that community college students often have complex schedules because of school, work, and family obligations; therefore, community colleges have difficulty creating the block schedules needed for learning communities. Visher et al. (2010) also mentioned that it can be challenging to find common threads between two classes. For example, students may have difficulty understanding how math and sociology relate to each other.

Various institutions have employed Supplemental Instruction (SI) to improve student success in mathematics. SI has been employed since the 1970s; however, the practice gained attention when Valencia Community College, an urban institution in Orlando, Florida, had a great deal of success with SI as part of the Achieving the Dream initiative (Phelps & Evans, 2006). SI is generally employed for courses, such as developmental math, which have high attrition and low success rates (Edlin & Guy, 2019).

SI consists of a session, in addition to the traditional class meeting, where students can gain extra practice and assistance and work collaboratively to solve problems. This may vary among institutions, but SI sessions are typically fifty minutes that meet twice a week (Maxwell, 1997; Phelps & Evans, 2006). An SI leader conducts such sessions. The SI leader is generally a former student or a tutor who has excelled in the course. The SI leader facilitates sessions where students can get extra assistance in the course. To best serve students, SI leaders often attend the main class to get a better feel for the instructor's pedagogy (Finney & Stoel, 2010). Like learning communities, SI sessions have been employed to combat isolation and to help students make connections (Phelps & Evans, 2006).

The impact of SI has been positive. Some math classes that have employed SI witnessed higher success rates than traditional classes (Edlin & Guy, 2019; Phelps & Evans, 2006; Price et al., 2012). However, success rates for SI are higher when students are required to attend the SI sessions as opposed to when these sessions are optional. Scheduling such sessions for community colleges can be challenging (Wright et al., 2002).

The use of SI parallels the corequisite model; however, there is one major difference. In the corequisite model, students must learn new material in the corequisite section and then apply it to the college-level course. SI sections generally provide extra review for the college-level courses.

Moving Forward

Underprepared students in higher education have existed since its inauguration. However, tighter funding and poor success rates in math have led to pressure from state legislatures to restructure and improve the mathematical model. The twenty-first century has seen various pedagogical initiatives aimed at improving math success rates. There have been both positives and negatives associated with such initiatives. However, to gain a better understanding of ways to thrive in community college math, it is imperative to learn about both the struggles and triumphs from students who have been successful.

References

Adams, P. (2017). Corequisite support case study: Community college of Baltimore. *Complete College America.* https://completecollege.org/article/corequisite-support-case-study-community-college-of-baltimore-county/

Altose, A. (2018). Embracing the value of college math. *Inside Higher ED.* https://www.insidehighered.com/views/2018/10/11/why-higher-ed-needs-new-approaches-teaching-math-opinion

American Council on Education (2004). *Putting college costs in context.* American Council on Education.

Arendale, D. (2002). *History of supplemental instruction (SI): Mainstreaming of developmental Education.* Center for Research on Developmental Education and Urban Literacy General College, University of Minnesota.

Arendale, D. (2003, October). Developmental education: Recognizing the past, preparing for the future. Paper presented at the Minnesota Association for Developmental Education 10th Annual Conference. Grand Rapids, MN.

Ariovich, L., & Walker, S. A. (2014). Assessing course redesign: The case of developmental math. *Research & Practice in Assessment, 49,* 45–57.

Ashburn, E. (2007). An $88-Million experiment to improve community colleges. *The Chronicle of Higher Education, 53*(33), A32.

Ashby, J., Sadera, W. A., & McNary, S. W. (2011). Comparing student success between 213 developmental math courses offered online, blended, and face-to-face. *Journal of Interactive Online Learning, 10*(3), 128–140.

Bahr, P. R. (2013). The deconstructive approach to understanding community college students' pathways and outcomes. *Community College Review, 41*(2), 137–153. https://doi.org/10.1177/0091552113486341

Bailey, T., Jeong, D. W., & Cho, S. W. (2010). Referral, enrollment, and completion in developmental education sequences in community colleges. *Economics of Education Review, 29*(2), 255–270.

Baldwin, J. (1975). Survey of developmental courses at colleges in the united states. *American Mathematics Association of Two-Year Colleges.*

Beamer, Z. (2020). Emporium developmental mathematics instruction: Standing at the threshold. *Journal of Developmental Education, 43*(2), 18–25.

Boylan, H. R. (2002). *What works: Research-based best practices in developmental education.* National Center for Developmental Education.

Boylan, H. R. (2008). Relentless leader's focus on developmental education: An interview with Byron McClenney. *Journal of Developmental Education, 31*(3), 16–18.

Boylan, H. R. (2011). Improving success in developmental mathematics: An interview with Paul Nolting. *Journal of Developmental Education, 34*(3), 12–41.

Boylan, H. R., & Bonham, B. S. (2007). 30 years of developmental education: A retrospective. *Journal of Developmental Education, 30*(3), 2–4.

Boylan, H. R., & White, W. G. (1987). Educating all the nation's people: The historical roots of developmental education part I. *Research in Developmental Education, 4*(4). 1–14.

Brier, E. (1986). Bridging the academic preparation gap: An historical overview. *Journal of Developmental Education, 8*(1), 2–5.

Brubacher, J. S. & Rudy, W. (1976). *Higher education in transition.* Harper & Row.

Bryk, A. S. (2015). *Learning to improve: How America's schools can get better at getting better*. Harvard Education Press.

Cafarella, B. (2020). Community college perspectives regarding Quantway. *Community College Journal of Research and Practice*. https://doi.org/10.1080/10668926.2020.1719940

Carnegie. (2020). *Quantway*. https://carnegiemathpathways.org/quantway/

Childers, A. B., & Lu, L. (2017). Computer based mastery learning in developmental math classrooms. *Journal of Developmental Education, 41*(1), 2–6, 8–9.

Clyburn, G. (2013). *Improving on the American dream: Math pathways to student success*. https://www.carnegiefoundation.org/wp-content/uploads/2013/09/Improving_on_the_American_Dream.pdf

Cohen, R., & Kelly, A. M. (2020). Mathematics as a factor in community college STEM performance, persistence, and degree attainment. *Journal of Research in Science Teaching 57*, 279–307. https://doi.org/10.1002/tea.21594

Dahlke, R. (2011). *How to succeed in college mathematics: A guide for the college mathematics student* (2nd ed.). BergWay Publishing.

Diaz, C. R. (2010). Transitions in developmental education: An interview with Rosemary Kerr. *Journal of Developmental Education, 34*(1), 20–25.

Edlin, M., & Guy, M. G. (2019). Mandatory and scheduled supplemental instruction in remedial algebra. *Journal of Developmental Education, 43*(1), 2–10.

Elrod, S. (2014) Quantitative reasoning: The next 'across the curriculum' movement. *Peer Review*. https://www.aacu.org/peerreview/2014/summer/elrod

Finney, J., & Stoel, C. F. (2010). Fostering student success: An interview with Julie Phelps. *Change: The Magazine of Higher Learning 42*(4), 38–43. National Center for Research on Teacher Learning. https://doi.org/10.1080/00091383.2010.489485

Francis, M., Wormington, S. V., & Hulleman, C. S. (2019). The costs of online learning: Examining tradeoffs between motivation and academic performance in community college math. *Frontiers in Psychology, 10*. https://www.frontiersin.org/articles/10.3389/fpsyg.2019.02054/full

Gordon, S. P. (2008). *What's wrong with college algebra? Primus, 18*(6), 516–541.

Goudas, A. M., & Boylan, H. (2012). Addressing flawed research in developmental education. *Journal of Developmental Education, 36*(1), 2–13.

Howard, L., & Whitaker, M. (2011). Unsuccessful and successful mathematics learning: Developmental students' perspectives. *Journal of Developmental Education 35*, 2–16.

Howington, H., Hartfield, T., & Hillyard, C. (2015). Faculty viewpoints on teaching Quantway. *Numeracy, 8*(1), 1–15. http://dx.doi.org/10.5038/1936-4660.8.1.10

Huang, M. (2018). *2016–2017 impact report: Six years of results from the Carnegie Math Pathways*. https://files.eric.ed.gov/fulltext/ED586608.pdf

Huang, M., & Yamada, H. (2017). Maintaining success rates: Does Statway sustain its impact as it scales to new classrooms and institutions? *Carnegie Foundations Math Pathways Technical Report*. Carnegie Foundation for the Advancement of Teaching.

Immerwahr, J., Friedman, W., & Ott, A. N. (2005). *Sharing the dream: How faculty, families and community leaders respond to community college reform* (Public Agenda). National Center for Research on Teacher Learning.

Jameson, M. M., & Fusco, B. R. (2013). Math anxiety, math self-concept, and math self-efficacy in adult learners compared to traditional undergraduate students. *Adult Education Quarterly, 64*(4), 306–322. https://doi.org/10.1177/0741713614541461

Kozakowski, W. (2019). Moving the classroom to the computer lab: Can online learning with in-person support improve outcomes in community colleges? *Economics of Education Review, 70,* 159–172. http://dx.doi.org/10.1016/j.econedurev.2019.03.004

Li, A. Y., Gandara, D., & Assalone, A. (2018). Equity or disparity: Do performance funding policies disadvantage 2-Year minority serving institutions? *Community College Review, 46*(3). https://doi.org/10.1177/0091552118778776

Mangan K. (2013). How Gates shapes higher-education policy. *Chronicle of Higher Education, 59(42), 14.*

Maxwell, M. (1997). *Improving student learning skills* (Rev. ed.). H & H Publishing.

Merseth, K. M. (2011). Update: Report on innovations in developmental mathematics–Moving mathematical graveyards. *Journal of Developmental Education, 34*(3), 32–39.

Olson, K. W. (1974). *The G.I. Bill, the veterans, and the colleges.* University Press of Kentucky.

Online Learning Consortium. (2018). *2018 annual report.* https://olc-wordpress-assets.s3.amazonaws.com/uploads/2019/04/OLC-2018-Annual-Report-Online.pdf

Pang, T. (2010). Improve math education, improve student retention. *The Chronicle of Higher Education, 56*(19), A30.

Park, T., Woods, C. S., Hu, S., & Jones, T. B. (2017). What happens to first-time-in college students when developmental education is optional? The case of developmental math and intermediate algebra in the first semester. *The Journal of Higher Education, 89*(1), 1–23. https://doi.org/10.1080/00221546.2017.1390970

Phelps, J. M., & Evans, R. (2006). Supplemental instruction in developmental mathematics (2006). *The Community College Enterprise, 12*(1). 21–37.

Phillip, A. (2011). The online equation. *Diverse Issues in Higher Education, 28*(3), 20.

Phillippe, K. A., & Sullivan, L. G. (2005). *National profile of community colleges: Trends & statistics* (4th ed.). Community College Press.

Price, J., Lumpkin, A. G., Seeman, E. A., & Bell, D. C. (2012). Evaluating the impact of supplemental instruction on short-and long-term retention of course content. *Journal of College Reading and Learning, 42*(2), 8–26. https://doi.org/10.1080/10790195.2012.10850352

Quarles, C. L., & Davis, M. (2017). Is learning in developmental math associated with community college outcomes? *Community College Review, 45*(1). https://doi.org/10.1177/0091552116673711.

Rizzo, J. M. (2006). State preferences for higher education spending: A panel data analysis, 229 1977-2001. In R. G. Ehrenberg (Ed.), *What's happening to public higher education? The shifting financial burden* (pp. 3–35). Johns Hopkins University Press.

Rutschow, E. Z., Schneider, E. (2011). Unlocking the gate: What we know about improving developmental education. *Building Knowledge to Improve Social Policy.* https://www.mdrc.org/sites/default/files/full_595.pdf

Smith, D., O'Hear, M. O., Baden, W., Hayden, D. Gordham, D., Ahuja, D., & Jacobsen, M. (1996). Factors influencing success in developmental mathematics: An Observational study. *Research and Teaching in Developmental Education, 13*(1), 33–43.

Spann, M. G. (1996). National center for developmental education: The formative years. *Journal of Developmental Education, 20*(2), 2–6.

State Higher Education Officers. (2010). *State higher education: State higher education finances,* FY 2010. Author.

Stigler, J. W., Givven, K. B., & Thompson, B. J. (2010). What community college developmental mathematics students understand about mathematics. *MathAMATYC Educator, 1*(3), 4–16. National Center for 232 Research on Teacher Learning.

Thelin, J. R. (2004). *A history of American higher education.* Johns Hopkins University Press.

Tinto, V. (1998). Colleges as communities: Taking research on student persistence seriously. *The Review of Higher Education, 21*(2), 167–177.

Twigg, C. A. (2011). The math emporium model: Higher education's silver bullet. *Change: The Magazine of Higher Learning, 43*(3), 25–34.

Vaughan, G. B. (1985). *The community college in America: A short history.* American Association of Community and Junior Colleges.

Visher, M. G., Schneider, E., Wathington, H., & Collado, H. (2010). *Scaling up learning communities: The experience of six community colleges.* National Center for Postsecondary Research, National Center for Research on Teacher Learning.

Wadsworth, L. M., Husman, J., Duggan, M. A., & Pennington, M. N. (2007). Online mathematics achievement: Effects of learning strategies and self-efficacy. *Journal of Developmental Education, 30*(3), 6–14.

Wang, X., Wang, Y., Wickersham, K., Sun, N., Chan, H. (2017). Math requirement fulfillment and educational success of community college students: A matter of when. *Community College Review, 45*(2), 99–118. https://doi.org/10.1177/0091552116682829

Weisbrod, B. A., Ballou, J. P., & Asch, E. D. (2008). *Mission and money: Understanding the university.* Cambridge University Press.

Weissman, E., Butcher, K. F., Schneider, E., Teres, J., Collado, H., Greenberg, D., & Welbeck, R. (2011). *Learning communities for students in developmental math: Impact studies at Queensborough and Houston community colleges.* National Center for Postsecondary Research, National Center for Research on Teacher Learning.

Wheeler, S. W., & Bray, N. (2017). Effective evaluation of developmental education: A mathematics example. *Journal of Developmental Education, 41*(1), 10–17.

Wheland, E., Konet, R. M., & Butler, K. (2003). Perceived inhibitors to mathematics success. *Journal of Developmental Education, 26*(3), 18–27.

Winerip, M. (2011, October 23). In college, working hard to learn high school material. *The New York Times.* https://www.nytimes.com/2011/10/24/education/24winerip.html

Wright, G. L., Wright, R. R., & Lamb, C. E. (2002). Developmental mathematics education and supplemental instruction: Pondering the potential. *Journal of Developmental Education, 26*(1), 30–35.

Xu, D., Dadgar, M. (2017). How effective are community colleges math courses for students with the lowest math skills? *Community College Review, 46*(1), 62–81. https://doi.org/10.1177/0091552117743789

Yamada, H., Bohannon, A. X., Grunow, A., & Thorn, C. A. (2018). Assessing the effectiveness of Quantway: A multilevel model with propensity score matching. *Community College Review, 46*(3), 257–287. https://doi.org/10.1177/0091552118771754

Yamada, H., & Bryk, A. S. (2016). Assessing the first two years' effectiveness of Statway: A multilevel model with propensity score matching. *Community College Review.* https://doi.org/10.1177/0091552116643162

3

The Study, Settings, and the Participants

Selecting the Participants

As a math professor at a large community college, I had the opportunity to interview students at my own institution who fit with the purpose of this study. However, I decided that given my position as a professor, this would not be appropriate. It is possible that one or more of the participants may enroll in one of my classes at a future time. Consequently, the student may feel uncomfortable with my knowing about his or her background after the interview. It is also possible that students may be recognized based on their responses if their college was identifiable. Accordingly, I decided to select my participants from outside community colleges where I had no affiliation.

My goal was to include at least twenty students in this study. That seemed to be a suitable number for a large qualitative study, as I hoped this size would generate enough data but would avoid repetition. However, to achieve balance in the study, I wanted a mix of males and females as well as a blend of ethnicities (white, black, Hispanic, etc.). I selected five community colleges and two midwestern universities, all in the United States. I included universities because such institutions may contain students suitable for my study.

The student services personnel helped me distribute the interview invitation to students. My invitation to participate included the following criteria for the study: The student had struggled in mathematics at some point and had generated some degree of math anxiety or dislike for math; the student placed into developmental math at a community college, and the student had received a grade of B or higher in a credit-bearing college-level math course (e.g., college algebra, quantitative reasoning, statistics, etc.) at a community college. I received twenty-eight responses, and I initially chose twenty participants from the five community colleges and two universities.

After the initial twenty interviews, I chose to conduct five more interviews. However, I found with each additional interview, I was learning less new information. Therefore, after twenty-five interviews, I reached satura-tion. This is the point where the researcher is no longer reaching new insights (Merriam, 2002b).

My selection of participants was an example of purposive sampling. According to Krathwohl (2009), "Purposive sampling is most often used in qualitative research to select those individuals or behaviors that will better inform the researcher regarding the current focus of the investigation" (p. 172). Patton (2002) added that purposive sampling encompasses delib-erately choosing participants who will add depth and understanding to a study; it is a method that includes seeking "information-rich" (p. 172) cases.

Conducting the Study

All interviews took place via Zoom conferencing with various follow-up questions via email. Because of the length of the interviews, all interviews were conducted in either two or three parts. The participants also com-pleted a demographic questionnaire, which consisted of short answers regarding their personal and academic backgrounds. I utilized standard open-ended questions for each interview. Patton (2002) has characterized this as a structured interview. Patton postulated that a structured interview keeps the interview in focus and efficiently utilizes both the interviewer's and the participants' time. There were lead questions, which were followed by what Kvale and Brinkmann (2009) referred to as probes, follow-up questions, and specifying questions to gain more depth to a response. To ensure accuracy, I had all interviews professionally transcribed.

This was a qualitative study; more specifically, this was a basic inter-pretive study. Merriam (2002b) asserted that in such a study, "the re-searcher is interested in understanding how participants make meaning of a situation or phenomena" (p. 6). More specifically, I sought to understand how this group of students interpreted their success in math coming into fruition. This study also resembled elements of a case study in that I was studying a group of students who shared the experience of struggles and success in community college math. Understandably, it can be difficult for participants to construct answers spontaneously. Therefore, I emailed each participant a copy of the lead questions several days in advance so that they could contemplate their answers.

I prioritized the imperative issues of informed consent and con-fidentiality. Kvale and Brinkmann (2009) stated that "Informed consent entails informing the research participants about the overall purpose of the investigation and the main features of the design, as well as of any possible risks and benefits from participation in the research project" (p. 70). My initial invitation explained the purpose of the study, my role as the interviewer, and their role as the participant. The participants signed a form consenting to partake in the study. According to Kvale, and Brinkmann, confidentiality "implies that private data identifying the participants will not be disclosed" (p. 72). I addressed confidentiality by assigning pseudonyms to all participants, community colleges, and universities in this study. Additionally, there were several times where the participants referenced an individual (e.g., teacher and counselor) by name, and in each case, I assigned a pseudonym.

Meet the Participants

In the next section, I will introduce and provide some brief background information about each participant. In subsequent chapters, I will provide more detailed information about the participants and their experiences in mathematics. I also asked each participant to provide a quote that represented their feelings regarding math as they entered community college.

Jessica

Jessica categorizes herself as a white female. She is a first-generation college student. Jessica's college endeavor began at the age of nineteen, in the fall semester of 2018, at Flores Community College, which is an urban community college with roughly 15,000 students. She placed into pre-algebra, which is three levels (courses) below college algebra. Jessica began college single, living with her parents, and working part-time as a server.

> *I tried to find a major that didn't include math. Guess what? That doesn't exist.*

Jessica completed her college-level math requirement at the end of the 2019 spring semester.

Robert

Robert identifies as a black male. He is a first-generation college student, and Robert's college journey began at the age of twenty-five, in the fall semester of 2016, at Drummond Community College, a suburban area with about 11,000 students. He tested into quantitative literacy, which is also known as Quantway 1, in the first semester of the Quantway sequence. Robert started community college as a married man working a full-time job driving a truck.

> *I wanted to go back to school for a few years, but I was scared, and math was a big reason for that.*

Robert completed his college-level math requirement at the conclusion of the spring semester of 2019.

Laura

Laura identifies as a Hispanic female. She is a second-generation college student, and her college endeavor began at the age of eighteen, in the fall semester of 2017, while attending Flores Community College. Laura placed into elementary algebra, which is two levels below college algebra. Laura began community college single, living with her parents, and working full time in a grocery store.

> *I wasn't really good at math, but I passed it in high school, so I was kind of mad that I had to take this developmental algebra before I could even take college math.*

Laura completed her college-level math requirement at the conclusion of the spring semester of 2019.

Dina

Dina identifies as a white female who is a first-generation college student. Her college career commenced at the age of thirty-five, in the fall semester of 2018, at Walsh Community College, which is in a suburban setting with roughly 11,000 students. Dina started community college divorced, with one child, living with her parents, and working part time in her family's restaurant. Dina placed into algebra 1, which is three levels below college algebra.

> *I don't know what scared me more, going back to school or taking math. I knew I needed more education, but every time I thought about going back to school, I couldn't breathe.*

Dina completed her college-level math requirement at the end of the spring semester of 2019.

Emily

Emily classifies as a white female who is a second-generation college student. Her college career began at the age of twenty, in the fall semester of 2016, at Arnold Community College, which is in an urban setting with about 17,000 students. Emily started community college single, living with her parents, and working part time as a cashier. Emily tested into arithmetic, a one-week refresher course that meets prior to the start of the semester. Students must pass this expediated course to even begin the traditional developmental algebra sequence. This course is four levels below college algebra.

I wanted a better life for myself. I knew that meant getting a college degree, but I hated math, and I couldn't believe I had to take the stupid math (arithmetic). It made me want to quit right away.

Emily completed her college-level math requirement at the end of the spring semester of 2019.

Joyce

Joyce classifies herself as a black female, and she is a first-generation college student. Her college journey commenced at the age of twenty-eight, in the fall semester of 2017, at Arnold Community College. Joyce placed into introduction to algebra, which is three levels below college algebra. Joyce started community college as a married woman with one son and working part time as a receptionist.

I wanted a better life for me and my kids, so I tried looking for a major with no math requirement. No luck. So, I picked human services because there wasn't much math.

Joyce completed her college-level math requirement at the conclusion of the spring semester, 2019.

Ron

Ron classifies himself a black male, and he is first-generation college student. Ron's college endeavor began at the age of nineteen, in the spring semester of 2018, at Walsh Community College. Ron tested into algebra 1, and he started college, single, living with his parents and working full time at a grocery store.

I didn't go to college after high school, and math was a big reason, but I figured out I needed more school, so I had to face my fears.

Ron successfully completed his college-level math requirement at the conclusion of the spring semester of 2019.

Andrea

Andrea identifies as a white female who is a first-generation college student. Andrea's college journey began at the age of eighteen, in the fall semester of 2017, at Drummond Community College, and she tested into introduction to algebra. Andrea started community college single, living with her parents, and working full time as a CVS associate.

I wasn't really afraid of math. I never liked it though. I was really ticked about being in a developmental math class when I passed algebra in high school.

Andrea satisfied her college-level math requirement at the conclusion of the summer term of 2019.

Mike

Mike categorizes himself as a white male. He is a first-generation college student. Mike's college career began at the age of forty, in the spring semester of 2017, at Blair Community College, which is in an urban area with 20,000 students. He tested into basic algebra, which is two levels below college algebra. Mike began his community college career as a married man with two children (ages twelve and ten) and working full time as a bank teller.

I was never good at school, and a lot of that was math. All those years, I still remember how stupid math made me feel, but I wanted to do better for my wife and kids, so I knew I had to go back to school.

Mike finished his basic college-level math requirement at the end of the fall semester of 2018.

At the time of the study, Mike was enrolled at Hollis University after transferring from Blair Community College. Hollis University is a liberal arts institution with roughly 12,000 students.

Tom

Tom classifies himself as a Hispanic male, who is a second-generation college student. Tom's college career began at the age of eighteen, in the fall semester of 2017, at Walsh Community College. He placed into algebra 2, which is two levels below college algebra. Tom began community college single, living with his parents, and not working.

I hated math, but I wasn't really scared of it-it was just boring. I never saw the point of it. I was so mad I had to take two remedial classes before college algebra. I even yelled at the people in the placement testing place. I told them they made a mistake, but it didn't do any good.

Tom successfully completed his college-level math requirement at the conclusion of the 2019 spring semester.

Larry

Larry identifies as a black male who is a first-generation college student. Larry began his college journey at the age of twenty, in the fall semester of

2017, at Flores Community College. He placed into pre-algebra, which is three levels below college algebra. Larry started his community college career single, living with his parents, and working part-time in a furniture store.

> *Just the thought of even enrolling in a math class made my heart beat so fast like I was having a panic attack.*

Larry completed his basic college-level math requirement at the conclusion of the fall semester of 2018. At the time of this study, Larry was attending Marcus University, a liberal arts institution with roughly 8,000 students.

Jerry

Jerry identifies as a white male, and he is a second-generation student. Jerry began his college career at Blair Community College, in the fall semester of 2016, at the age of twenty-five, and he placed into basic algebra. Jerry began community college married and working part-time as a mail carrier.

> *It actually made me laugh. I had to pass two math classes just to reach a college-level math class. Yeah and unicorns are real!*

Jerry completed his college-level requirement at end of the spring semester of 2019.

Kathryn

Kathryn classifies herself as a white female who is a first-generation college student. Kathryn's college career commenced at the age of eighteen, in the fall semester of 2016, at Drummond Community College, and she placed into introduction to algebra. Kathryn started her community college career single, living with her parents, and working part-time as a server.

> *I don't know if I was afraid of math; it just never made any sense to me. It was always like a foreign language. Now more of this in college. Great!*

Kathryn completed her college-level math requirement at the end of the summer term of 2019.

Deb

Deb identifies as a black female, and she is a first-generation college student. Deb started her college career at Arnold Community College at the age of eighteen, at the start of spring semester of 2018. She placed into introduction to algebra. Deb started community college single, living with her parents, and working part-time as a receptionist in a doctor's office.

*I admit it; I was looking for a degree where I could get a job, but more im-
portantly I didn't want to take math. In fact, I first wanted to only go for a two-
year nursing degree where you don't have to pass college math, but people kept
telling me I needed the four-year degree. I just didn't want to take math.*

Deb completed her introductory college-level math requirement at the
conclusion of the spring semester of 2019.

Denise

Denise classifies herself as a Hispanic female, who is a first-generation
college student. Denise began her college career at Blair Community College
at the age of twenty-seven, at the start of the spring semester 2018, and she
tested into basic algebra. Denise began her community college single (divorced), living on her own, and working full time as a technician in a
doctor's office.

I seriously cried as I was registering for my math class. Enough said.

Denise completed her college-level math requirement at the end of the
spring term in 2019.

Tara

Tara identifies as an Indian female, and she is a second-generation college
student. Tara began her college career at Walsh Community College at the
age of twenty-one, at the start of fall semester 2016, and placed into algebra
1. Tara started community college single, living with her parents, and
working part-time at a day care center.

*As school got closer, I would think about taking math, and I would have to run
to the bathroom and throw up; I'm serious.*

Tara successfully completed her college-level math requirement at the
conclusion of spring semester of 2019.

Harold

Harold categories himself as a white male, and he is a first-generation
college student. Harold's college career began at Drummond College, at the
start of the fall semester 2018, at the age of nineteen, and he placed into
intermediate algebra, which is one level below college algebra. Harold
started community college single, living with his parents, and not working.

*What did I remember about math? There were fractions, factoring, something
about factorials. Yes, lots of f words.*

Harold finished his college-level math requirement at the conclusion of the summer term of 2019.

Otis

Otis identifies as a white male, and he is a second-generation college student. Otis began his college career at Arnold Community College at the age of forty-eight at the start of the spring semester of 2018. Otis began community college as a married father of three (ages eighteen, fifteen, and ten) and not working while collecting disability. Like Emily, Otis tested into the arithmetic refresher.

> *I can still feel my heart beating through my chest when I thought about taking math. I was always bad at it, but it had been thirty years since I was even in school.*

Otis completed his college-level math requirement at the end of the spring semester of 2019.

Dan

Dan identifies as a white male, and he is a first-generation college student. Dan's college career commenced in the fall semester of 2016 at Flores Community College at the age of thirty-five, and he placed into pre-algebra. Dan started community college as a married man with one child (four months) and working part time for a landscaping company.

> *I was excited about going back to school. I knew it was the right thing to do, but math always made me feel so stupid. I had thought about going back to school before, but I didn't because of math.*

Dan completed his college-level math requirement at the conclusion of the spring semester of 2019.

Todd

Todd classifies himself as a white male who is a third-generation college student. Todd began his career at the start of the fall 2017 semester at Flores Community College at the age of eighteen, and he tested into pre-algebra. Todd started community college single, living with his parents, and working a part-time job as a receptionist at a hotel.

> *I wanted to go into the hotel business with my dad. I thought, No way is there math for this degree. I almost fell down when I found out there was.*

Todd completed his college-level math requirement at the end of the spring semester of 2019.

Joe

Joe identifies as a black male, and he is a first-generation college student. Joe's college career began at the start of the fall 2015 semester at Blair Community College at the age of eighteen, and he tested into basic algebra. Joe began his community college career single, living with his parents, and working a part-time job at McDonald's.

> *My parents kinda made me go to college; I didn't want to. I always hated school, and math was a big reason. I just never saw the point of learning math.*

Joe completed his college-level math requirement at the end of the spring 2019 semester.

Rosemary

Rosemary identifies as a white female who is a third-generation college student. Rosemary's college career commenced at the start of the fall term of 2016 at Drummond Community College at the age of twenty-nine and tested into quantitative literacy. Rosemary started community college as a single woman, living on her own, and collecting unemployment.

> *I didn't even remember any math I had ever learned. All I remembered was that it scared me and made me cry a lot. I put off going back to school because of taking math.*

Rosemary completed her college-level math requirement at the end of the spring semester of 2019.

Cindy

Cindy classifies herself as a Hispanic female, and she is a first-generation college student. Cindy's college career began at the start of the fall semester of 2017 at Flores Community College at the age of nineteen, and she tested into pre-algebra. Cindy started community college single, living with her parents, and working part-time as a veterinary technician. Cindy completed her college-level math requirement at the conclusion of the spring semester of 2019.

> *Three math classes just to take college math? Like I'll still be there to even take college math!*

Adam

Adam identifies as a black male who is a second-generation college student. Adam began his college career at the start of the fall 2017 semester at Drummond Community College at the age of eighteen and tested into introduction to algebra. Adam began community college single, living with his parents, and not working.

I don't know why I'm even going; math is gonna kill me.

Adam completed his college-level math requirement at the end of spring semester 2019.

Audrey

Audrey identifies as a white female who is a second-generation college student. She began her college endeavor in the fall semester of 2018 at Arnold Community College at the age of twenty-five and placed into introduction to algebra. Audrey began community college living with her mother and not working.

I was scared, but I knew I had to deal with math.

Audrey completed her college-level math requirement at the end of the summer term of 2019.

I also asked the participants to list their majors when enrolling in college. Jessica and Jerry enrolled in a criminal justice program. Laura, Andrea, Kathryn, Harold, Emily, and Cindy enrolled in early childhood programs. Robert, Joyce, Tara, Rosemary, Audrey, and Otis enrolled in a human services degree program. Todd enrolled in a hospitality program. Deb and Denise enrolled in nursing programs; however, they intended to pursue four-year degrees in nursing, which requires successful completion of college-level math. Dina enrolled in a fine arts degree. Tom joined an aviation program. Dan, Larry, Joe, Audrey, Ron, Mike, and Adam enrolled in general liberal arts degrees. Jessica, Joe, Tara, Emily, Cindy, Joyce, Deb, Dan, Mike, and Adam specified that they chose their initial degrees because such degrees required minimal math requirements.

Introduction to algebra and pre-algebra are the lowest level developmental courses at Drummond Community College and Flores Community College, respectively. Algebra 1 is the lowest level developmental math course at Walsh Community College. Quantway 1 (quantitative literacy) and Statway 1 are the lowest level math courses at Drummond Community College for students embarking on the Quantway and Statway pathways. Students who score low on the placement exam are referred to an external state-funded adult basic education program. Arnold Community College offer a week-long arithmetic refresher course for students who test below basic algebra and introduction to algebra, respectively. While several of the participants did not attempt college algebra, I wanted to convey the length of the traditional pathway to college algebra. Therefore, I noted how many levels their starting course was below college algebra.

All the aforementioned math courses, with the exception of the arithmetic refresher, allow the use of a calculator. The pre-algebra courses, the introduction to algebra, and basic algebra courses allow a four-function calculator, and the elementary algebra and intermediate algebra courses permit a scientific calculator.

Summary

As a requirement for the study, all students tested into some type of developmental math. All participants approached their college endeavor with some sort of disdain and dread regarding attempting mathematics. In the next chapter, the students will provide in-depth details as to what led to their negative feelings toward math. In many cases, there were situations that only added to the mindsets of contempt and trepidation.

References

Krathwohl, D. R. (2009). *Methods of education and social science research* (3rd ed.). Long Grove, IL: Waveland Press, Inc.

Kvale, S., & Brinkmann, S. (2009). *Interviews: Learning the craft of qualitative research interviewing* (2nd ed.). Sage Publications, Inc.

Merriam, S. B. (2002a). Assessing and evaluating qualitative research. In S. B. Merriam and Associates (Eds.), *Qualitative research in practice: Examples for discussion and analysis* (pp. 18–33). Jossey-Bass.

Merriam, S. B. (2002b). Introduction to qualitative research. In S. B. Merriam and Associates (Eds.), *Qualitative research in practice: Examples for discussion and analysis* (pp. 3–17). Jossey-Bass.

Patton, M. Q. (2002). *Qualitative research & evaluation methods* (3rd ed.). Sage Publications, Inc.

Ridenour, C., & Newman, I. (2008). *Mixed methods research: Exploring the interactive continuum.* Southern Illinois University Press.

4

Prior Experiences in Math

Several students indicated that they viewed mathematics as an obstacle to their potential success in higher education. How can one discipline cause so much contempt and panic for students, even before setting foot on a college campus? What experiences could lead to such animosity toward one subject? Several of the participants shared experiences from elementary and secondary education that developed and embellished their dislike of mathematics.

The Positives

A few of the participants relayed some encouraging early experiences with mathematics.

Denise shared:

> I have some nice memories of learning math early. I remember my fourth grade teacher made math fun. We would play all sorts of games, learning difference shapes. Even when we learned multiplication, she made it fun. She would come around and help us, and she was kind to us if we didn't get it. She would even joke around and get us to laugh at ourselves and our mistakes.

Todd added:

> I don't remember too much, but I remember I was good at things like adding, subtracting, and multiplying. It just made sense to me. My teachers used to have me help other kids who struggled, and that always made me feel good.

Cindy reflected:

> Up until about fifth grade, I always got As in math. I never really thought about it. It just came easy to me.

Jessica recalled:

> Math was fun. I remember learning math using those blocks [base ten blocks]. I remember working together with my classmates and my teachers supporting us. I always did well.

Rosemary called to mind:

> *Learning to borrow in subtraction, like 42 minus 19, was really scary, but my teacher had a way of making it easy. We used these blocks, and I could understand what I was doing. I was so excited. I even showed my mom how to do it that night.*

Audrey shared:

> *I was a good student in elementary school; I can't say I liked math or really remember a whole lot, but I did pretty good in all my classes.*

Mixed Feelings

> *Sometimes I was good at math [in elementary school]; sometimes I wasn't. I did good in multiplication and division. I liked geometry, but I hated word problems. I could never understand what I was supposed to do. I didn't get fractions at all,* recollected Joe.

> *In grade school, math was either my favorite subject or my least favorite subject. Till about third or fourth grade, I was great at it. As soon as we got to fractions, I was lost, especially canceling fractions. Mixed numbers made no sense to me either. I remember my fifth grade teacher yelling at me that I didn't listen. Maybe I didn't. I don't know,* said Robert.

Cindy shared her thoughts:

> *Math was fun till it wasn't. We got to fifth grade and got into geometry with angles and all kinds of confusing stuff. The word problems also got harder. When it was more basic, it was easy. It was fun to do activities and games with shapes like rectangles and triangles, but when we had to do word problems with them and use formulas, it made less sense.*

Joyce recalled:

> *I don't really remember the stuff we learned in grade school math; it was more about the teachers. When I had teachers, who were patient and nice, I did pretty well, but then I had some teachers who just yelled when I didn't understand stuff and made me feel stupid.*

Larry remembered:

> *In third grade, my teacher actually called me "Mr. Expert in Math". I knew all my multiplication tables and I was good at pretty much everything. That all changed with long division in I think it was fourth grade. It was so confusing.*

I remember starting to panic because it was coming so easy for other kids, but I didn't get it. That first step when you are supposed to guess how many times a number goes into another made no sense. I remember my teacher kept yelling, "Trial your answer!" I didn't know what she was talking about.

Deb summarized:

Look, I honestly don't remember too much about grade school, but I was talking with my mom, actually, just before this interview. She [Deb's mother] said I learned my multiplication tables really easy, but I couldn't get division. My parents couldn't understand how I could multiply so easily but division was so hard. She said, "You could do nine times three, but you couldn't do nine divided by three". She said I was never the same in math after that.

Mike synopsized:

I was OK when I could memorize how to do stuff. For example, with fractions, I memorized how to multiply and divide fractions OK, but I could not, for the life of me, understand how to cancel, you know cross cancel when you multiply? I guess because I had to actually understand it instead of memorizing it. I remember I would multiply fractions and get answers like 27 over 45, and my teacher would yell, "You don't pay attention to what I tell you!"

Tara recalled:

It really depended on the teacher. When I had a nice teacher, I liked math, but some teachers were really mean and would yell, and that made me scared of math.

The Negatives

I can't remember the topic but this one time in, I think it was fourth grade, my teacher called me to the board to do a math problem, and then she left the room. I couldn't do it. I stood there looking at these numbers with all the kids laughing at me. When she came back in, she yelled at me. I wanted to crawl into a hole, recalled Jerry.

Otis recollected:

I was never good at math; it was always a struggle. Lots of frustration in my house with my parents with me trying to my homework because I just didn't get or want to do math.

Otis paused and then continued:

> *Here is something that pretty much symbolizes my math experiences. Early on,*
> *I just gave up, and most of the time I would just write any old answer down in*
> *my homework to look like I did my homework. So, this one time in fourth grade,*
> *we were going over homework, and she [the teacher] called on me. I can't believe*
> *I remember this, but I wrote 5 down for the answer, and yeah, it was a random*
> *guess. I sheepishly said, "I think I did this wrong". She looked at my answer*
> *and tore into me for what seemed like forever. I just remember her saying, "How*
> *stupid of you! How stupid of you! You're stupid!"*

I asked Otis if he told his parents what this teacher said, and he replied,

> *No, I thought I would just get into more trouble.*

Emily said,

> *I don't remember what we did in grade school math, but I was bad at it, and I*
> *even had to go to summer school once or twice. I remember something else. We*
> *used to go over our math answers in a round-robin way. You know, that's when*
> *the teacher goes around the room in order and everyone has to answer. I still*
> *remember my heart pounding when it was almost my turn. I had the wrong*
> *answer, and everyone was going to find out.*

Laura shared:

> *Math was just so boring. I used to dread my teachers saying, "Take out your*
> *math books."I hated it all. Numbers just made no sense to me.*

Dina conveyed:

> *I can't even tell you what we did in math, but I was always behind. It's like I*
> *couldn't keep up. Do you know what it feels like to just want to understand*
> *what the teacher is talking about, but you just can't?*

Simple Concepts Were No Longer Simple

A few participants recalled that their math anxiety increased as the content became more difficult. Cindy recalled:

> *Around middle school, stuff that used to be easy wasn't so easy. I remember*
> *learning about sets, you know the stuff with intersections, unions, and Venn*
> *diagrams? In fifth grade, I was really good at that. In middle school, we started*

doing other stuff that involved sets, but it was so much harder. I don't re-
member what it was, but I missed how easy sets used to be.

Dan concurred:

I think it was eighth grade. My teacher said, "Today we are going to learn about
circles", and I was thinking, "Oh yeah, I got this". I knew the formulas
for circumference of a circle and area of a circle. Then, all the sudden he starts
talking about degrees and arcs and angles, and I got lost.

Jessica added:

In elementary school, I was pretty good at word problems. I mean; all you really
had to do was figure out if you had to add, subtract, multiply, or divide. You
know, just look for the key words. Man, they [word problems] really blew my
mind in algebra. That's when we had to let x equal this or let y equal that. I
remember the stuff about finding consecutive odd integers or consecutive even
integers. It just didn't make sense anymore.

Kathryn recalled:

I remember in high school, our teacher asked us if we remembered how to do
fractions, and she even went over an example. I hated fractions, but I was
like, "OK, I remember this". Then, she started doing fractions with x's and
y's and all the algebra stuff [operations with rational expressions]. No way. I
was lost.

Math Became a Foreign Language

A few participants articulated that the discipline of math began to resemble
a foreign language as the content became more arduous. Denise elaborated:

It was in middle school and we were doing probability. My teacher was talking
about "with replacement and without replacement", and I'm thinking, "What
exactly are we replacing? What is she talking about?" That is just one example
of what made me hate math. There were so many times like that I felt so
helplessly lost.

Harold called to mind:

I think it was eighth grade, and the teacher was doing this stuff with all these
explanation points, and I'm thinking, "This is a math class, not an English

class, right?" OK, later I realized those were called factorials, but still I think that was the moment I realized math made no sense to me at all whatsoever.

Kathryn shared:

I feel like this happened many times, but this one time in tenth grade all the sudden we started doing logic; you know the stuff with p and q and the truth tables? I never felt so lost in my life. Someone even asked the teacher, "What does this have to do with math?" All he said was, "Math is logic". OK, thanks! But when you are sitting in a room feeling so lost and confused, it makes you nervous and scared. This was a big reason I hated math. Knowing I will feel confused and stupid and that I might get called on and embarrassed scared me.

Andrea questioned:

Why was a negative number times a negative a positive? Why wasn't the answer like a really big negative? I was thinking, "Who comes up with this crazy stuff?"

Emily recalled:

It was just basic terminology that made no sense so many times. I think we started doing angles in fourth or fifth grade, but I never got it. One time in seventh grade my teacher called on me, and I didn't know the answer to any of her questions. She finally said, "Do you even know what an angle is?" I said no. She shook her head and just moved on.

Mike recalled:

It's probably weird I remember this, but I'll tell you why. This one time in eighth grade, we were doing graphs. I don't remember the exact equation, but let's say it was something like $3x + 4y = 12$. We needed to graph the equation, and my teacher just substituted 0 for x and 0 for y. I didn't get where she came up with the zeros. It was literally the only time I ever raised my hand to ask a question. I asked, "Where did you get the 0?" She said, "I made it up", and she moved on. I was thinking, "So that's math? We just make up numbers?" All these years; that just stood out.

Cindy shared an ambiguous memory from middle school:

My teacher was talking about dividing numbers by zero. She showed us eight divided by zero and just said, "That is undefined; mathematicians don't know what that means". I was like, "OK, so why is it undefined? Are mathematicians just dumb?" Stuff like that just made me feel more lost.

High School Completion of Mathematics

I asked all participants how many years of high school mathematics they completed as well as the levels of mathematics. For all participants, except for Otis, Dina, and Mike, their high schools required students to complete two years of algebra (algebra 1 and algebra 2) along with one year of geometry. The sequence consisted of completing algebra 1 freshman year, geometry sophomore year, and algebra 2 junior year. Otis needed to complete only two years of any math. He completed a basic math class the first year of high school and a business math class his sophomore year. Dina completed one year of basic math and another of basic algebra. Mike completed three years of math; however, his courses consisted of a basic math class, a basic algebra class, and a consumer (business) math class.

The requirements for two years of algebra and one year of geometry were adaptable. The participants' high schools employed a tracking system. While their various high schools labeled the math tracks differently, the themes were similar. Honors algebra and geometry were the highest track; however, none of the participants attempted this track. Next, there was the average track, or college preparatory track, for students who were on the path to college. Behind that, there was the below-average track. For example, the below-average track for algebra 1 contains less content and is slower paced than the average algebra 1. Students could also opt out of algebra 2 and enroll in a career-based math course.

Most participants dwelled in the lower level math throughout high school. Harold was the one exception; he completed the average-level sequence of algebra and geometry. Cindy, Laura, Denise, and Kathryn began their high school careers in the average-level math but were demoted to the lower level at some point. Emily, Joyce, Ron, Tara, Larry, Deb, Dan, Joe, and Adam opted to complete the career-based math junior year instead of algebra 2. Audrey dropped out of high school after her junior year. I requested that the participants share some of their experiences from high school math, the years leading up to their community college math experience.

Mind the Gaps

Some participants mentioned that the gaps in their mathematical knowledge base were a source of frustration, especially in high school.

> *I still had trouble with multiplication and division without the calculator. When you can't multiply or divide in your head, it keeps you from doing so many things,* said Emily.

The signs killed me. There were so many times I was starting to under-stand something in math. I'm like, I got this, and then bam! Those freakin' signs, recalled Adam.

If I saw we had to do fractions, I was lost; that was it. My teacher wrote a fraction on the board; my head hit the desk, shared Joe.

Cindy shared:

Tenth grade geometry was a nightmare. That's when you have all those proofs. I never got the basic stuff with angles and rays and line segments and vertex and all that. I didn't even understand what a ray was! So, there was no way I was gonna get all those complicated proofs. It was so embarrassing. I felt so stupid.

Ron recalled:

I hated all the math vocabulary. Anytime my teachers would talk about coef-ficients, factors, evaluate, expressions, and all that, I just felt even more lost.

Kathryn reflected:

I think it was seventh or eighth grade we learned about radicals, you know like taking the square root of 50 or the square of 80? I never understood it. I mean I could take the square root of 50 on my calculator, but I didn't know how to do it without the calculator. So, when I got ninth grade, we did those radicals but this time they had x's and y's. I'm like if I can't do this with plain numbers, I can't do it with letters!

Difficulty Keeping Pace

Some participants discussed the challenge of keeping pace with the flow of high school mathematics. Cindy elaborated:

I started in the college prep math (average level), and I barely passed. There was some algebra that made sense to me, but most of it didn't. Then, I started geometry, and I was totally lost, so they got my butt out of there. I had been struggling in math for a while, so I was like "whatever". I barely passed the rest of high school math. I didn't get the rest of algebra or geometry. I think my teachers just passed me even though I didn't really pass, if you know what I'm saying.

Denise recalled:

> *In the mainstream (average level), I just couldn't keep up. I was so lost, but I guess I didn't try that hard.*

Laura reflected:

> *I just couldn't keep up with the class (average-level algebra 1) so they kicked me out. I feel like maybe I could have done better if it moved slower. I did OK in the fundamental (lower level) math, but there was still a lot of stuff I didn't get.*

Todd got straight to the point:

> *It was simple. The more math I took, the less I understood, and the dumber I felt.*

Audrey recalled:

> *I started falling behind in math in middle school. It's like I would get confused and not get one thing, and then I couldn't get another thing because I didn't get the first thing, if that makes sense. But for me, I was dealing with a lot of depression and anxiety, and I couldn't focus on anything.*

Dina reflected:

> *I was the stupid girl who could never keep up. It seemed like my teachers just gave up on me. They would go from rolling their eyes that I couldn't get it and then would just stop calling on me. By high school, I just sat by myself not getting it. I don't know how I passed. I guess they [Dina's teachers] felt sorry for me.*

Tara shared:

> *I used to cry and throw up before math class. Lots of reasons, but it was just that feeling that I knew I was going to be lost. I knew whatever the teacher said, I wouldn't get it.*

Humiliation 101

A few of the participants articulated that math was simply a time for anxiety and embarrassment.

My answers in class were so stupid that the other kids would laugh at me and call me names. This one time the teacher called on me, and I didn't know the answer. He kept asking me questions, and I just froze. The other kids were laughing so much that I just ran out of the room crying, shared Tara.

Larry conveyed his experience:

I got laughed at a lot in math class. We used to go around the room, and the teacher would call on us. By the time it was my turn, the kids were all snickering and laughing because they knew my answer was going to be stupid.

Deb recalled,

Looking at all that algebra used to make feel so dumb. I just used to stare at those problems, and I just couldn't do anything. I don't know how I passed; I guess I got credit for being there.

Mike said,

I was in the "stupid math", and I even had to go to summer school for "stupid math". I hated math; I was scared of math; I just didn't get math.

Audrey reflected:

I fell so far behind, and I felt so dumb in math. I remember this one time my teacher called on me, and I got the answer wrong, and she kept asking me questions, and I kept getting them wrong because I was like totally lost. I mean I had no idea what we were doing. Every time she asked me a question, the other kids would laugh louder and louder. After a while, I started crying.

Audrey also received some very unwelcome attention from a male math teacher:

This one time this guy [Audrey's teacher] made me stand up from my seat, and he said, "How can you be so beautiful yet so stupid?" That was the last time I was ever in a math class.

Harold exclaimed:

Math should have been called humiliation 101. The other kids used to laugh at my answers in class all the time.

Personal Problems, Poor Learning Environments, and Apathy

I went to school in the hood, so there wasn't a whole lot of learning, if you know what I mean. My teachers couldn't control the class. Kids were always talking and messing around. I remember more teachers just yelling at us to be quiet and stay in our seats, not really teaching us stuff. I didn't understand the math a whole lot, but my teachers gave us a lot of help on tests; you know they would pretty much give us the answers, said Joe.

My math classes were out of control recalled Joyce. *Kids talking, throwing stuff, one time this guy pulled a girl's pants down when the teacher was talking. It was hard to learn anything.*

Kathryn recalled:

In the college-prep class, the kids were a little more serious about learning. In the applied classes, my teachers couldn't control the class, so there wasn't a lot of teaching going on. My teachers used to show us a lot of videos of people teaching math. We had a lot of free time to talk during class too. I feel like I had to teach myself a lot. I still didn't understand algebra and geometry, even in the applied class. To be honest with you, by high school, I just didn't care about school.

Emily shared:

I had a lot of problems in high school; I struggled with anxiety and depression, and I missed a lot of school. I was in special education, so I took my tests in a resource room with another teacher who helped me. She would pretty much take the tests for me, so that's how I passed math.

Adam recalled:

I got into a bad crowd in high school. I used to cut class a lot and hang out with my friends. I hated math; in [math] class, it felt like my teacher was speaking another language. I never studied; I failed all my tests. I guess skipping so many classes didn't help.

Robert shared:

High school was a miserable time. I was depressed a lot, and I didn't care about my schoolwork. I will say, being so bad at math just gave me one more reason to feel worse about myself.

Rosemary confessed:

I admit it; I didn't study; I didn't know how to study. It just got to where I was so far behind in math, I didn't know how to fix it.

Jessica recalled:

> *I started off liking math, but then as it got harder, I got more afraid of it. Now, as I got into high school, I just stopped caring, because I just didn't get it. I had to take math my senior year, and I was failing. I needed an 85% on my math final to pass the class and graduate. I was never so scared in my life. My mom got me a tutor, and I studied so hard. I guess I did well enough to pass. I never wanted to take math again ever.*

Jerry said,

> *I cut class all the time; I didn't do any work. I just got so bad at school; I stopped caring. You know how I passed? I got special services and people helped me with all my tests. Without that, I would have failed.*

Math as Compared to Other Subjects

The students clearly detested and often struggled extensively in math; however, I asked how they performed in math compared to their other subjects. None of the participants scored higher than a C (70–79%) for their cumulative average in math. Jessica, Emily, Joyce, Larry, Jerry, Deb, Otis, Tara, Rosemary, and Adam achieved a D (65–69%) average. Several of the participants stated that math was their worst subject.

> *I did all right in English, in social studies, but I was always failing math*, said Jessica.

> *The thing about social studies is you can always look up the answers in the book; you can memorize facts and days; you can't do that in math*, asserted Rosemary.

> *I was interested in books; I liked English*, recalled Harold. *I liked writing too. I like how there isn't one right answer in writing. That's one thing that makes math so scary; there is only one right answer.*

Dina shared,

> *I'm an artist, so I did well in my art classes. I guess classes where I could be creative I did well, but not math.*

Adam remembered,

> *I didn't do too good in any class, except for easy classes like gym and stuff, but yeah, math and science was always the worst for me.*

Summary

Overall, the math experiences for the participants during their K-12 career were negative. Some of the students had positive math encounters early in their school career; however, the combination of intimidating teachers and the inability to keep pace with increasingly challenging content quickly deteriorated their outlook for math. By high school, the students' emotions ranged from anxiety to disdain to apathy regarding math. Poor learning environments, too many gaps in their math knowledge base, poor study habits, and personal barriers further lowered their self-efficacy toward math. The participants then had to approach this hated discipline in the unfamiliar confines of a community college setting.

5

Attempting Math and Community College

At various stages of their lives, the participants chose to continue their education at assorted community colleges. Several of the students were attempting to create a better life for themselves by furthering their education; however, math remained a major barrier. In order to earn college credit in math, the students had to successfully complete their developmental math requirements.

The students eventually took varying pathways to complete their college-level math requirements. However, there were two commonalities among all the participants: each student had to begin the endeavor by successfully completing at least one stand-alone developmental math course, and all participants attempted developmental math during their first semester.

Starting Community College

The regular major league baseball season is six months (April through September), and in that time teams compete to win their division to advance into the playoffs. There is common expression that "you can't win a division in April, but you can lose it." More specifically a bad start to the season by losing too many games can put a team too far behind and make winning a division very difficult or impossible. The participants were attempting to complete their least favorite and, in several cases, most terrifying subject. However, they were also endeavoring this arduous task in a new and unfamiliar environment. A good start was imperative.h Therefore, I asked the participants their feelings regarding starting this new experience and their approach to their higher education career.

Flores Community College

Jessica and Larry struggled in the workforce directly after high school, but both participants faced anxiety regarding returning to school. Jessica shared:

Going to college was scary. I didn't know anyone, and I knew I was a bad student. I took a year off after high school because I was so sick of school. I got my own apartment and I was working as a waitress at Chili's, and it really, really sucked. I hated my job, and I could hardly pay my bills. I was like, I don't want to do this for the rest of my life.

So, I decided to go back to school. I wanted to do well, but I just didn't know how to even be a good student.

Larry elaborated:

I didn't want to go to school. I started working at copy shop after high school, but I got laid off. I spent a few months trying to get more work, but I could only get part-time jobs. It was a bad time. So, I knew I needed to get more education. I mean, I had no skills. I was scared about going back to school though. I was a bad student, and I was bad at math. I was wondering if people were gonna pick on me like they did in high school.

Laura's mother gave her an ultimatum:

My mom told me I either had to go to college or she was going to charge me rent for living at home. She's a single mom, and I got two younger brothers. Some of my girls from high school were going to Flores, so I knew some people. I wasn't that nervous. I just wasn't looking forward to more school.

Dan recalled:

I knew I needed to go back to school, but I was scared. I had a wife and a kid, and it felt like there already wasn't enough time in the day. How was I gonna go to school, work, study and do homework and still be there for my family? Oh yeah, and I hadn't done math in seventeen years, and I was bad at it then!

Todd questioned:

I knew what I wanted to do. I wanted to join the family hotel business, and that meant more school, but more math? Why the hell did I need to know how to do equations and stuff to run a hotel?

Cindy shared:

I knew a lot of the same people from kindergarten all the way to twelfth grade. Now I'm going to this place where I don't know nobody. I'm shy; I don't talk to people until they talk to me.

Flores Community College (FCC) offers an optional head start program for incoming students who are enrolled in pre-algebra. The head start program meets for five days for four hours per day, the week prior to the start of the semester, and it is staffed with three math faculty members and math student-tutors. In the head start program, students work independently, in a lab, on the inaugural pre-algebra material, and they can get one-on-one assistance. There are pre-prepared problems for all students on MyLab Math, and students can work as far as they are able to. The completion of this work does not count toward the pre-algebra course; it is simply a way for students to sharpen their skills and get a head start on the course content. Throughout the head start program, the math faculty also conduct workshops on basic study skills and provide tips for math success. Jessica, Larry, and Cindy opted to participate in the head start program. Dan and Todd decided not to attend.

> *My parents wanted me to go [to the head start class], but I was like, no way; what's in it for me? Let's see, I could work and make money or do math for free,* recalled Todd.

> *When I saw I wasn't getting any credit or anything, I was like, no I'm not giving up a week just to do math,* said Dan.

I asked the participants who attended about their experience in the head start program.

Jessica reflected:

> *It was good. I finally understood what to do with fractions and decimals and even the signs [signed numbers]. It was nice to work with people and get individual attention. I was still freaked out about going to the actual class, but I'm glad I went to the head start thing.*

Larry shared:

> *Oh, the program was a big help. Someone finally got me to understand fractions and stuff, you know like adding, multiplying, and reducing and all that. I even started understanding those signed number rules at the end of the week. I was thinking if these people could explain it to me, maybe I could actually learn something here, but I was still scared about starting school. You know what? I decided I was gonna do whatever it took to pass. Whatever my professors wanted me to do, I was gonna do. I was gonna work as hard as I needed to.*

However, Larry had some concerns after attending head start.

> *One thing I didn't like was these cliques that started forming; you know, people started making friends and hanging out with each other, and I felt left out. You know; it was like high school. Cliques just made me uncomfortable.*

Cindy recalled:

> *I loved it! I got to meet some new people, which can be hard for me. The teachers and the tutors were awesome! I did some work on fractions and decimals and the signs. I kinda remembered how to do that stuff, but it was real good to get the review. It was nice to meet some of the math teachers at Flores too. They seemed real nice and maybe this place wouldn't be so scary.*

Drummond Community College

Robert recalled:

> *I don't remember a whole lot about how I felt going to college. I was struggling with depression a lot, so I was in a funk most of the time. That's why I picked human services; I thought I could help other people who were depressed. I just knew I had been in and out jobs for years, and I needed a better life for me and my wife.*

Andrea conveyed:

> *I was kinda sad about leaving my friends in high school, but I'm pretty outgoing, so I knew I would meet new people pretty fast. I was an OK student in high school, so I wasn't too nervous about new classes.*

Kathryn shared:

> *I really didn't want to keep going to school, but I knew a lot of people who didn't go to college, and they wound up working these really crappy jobs they hate for no money, so I thought I better get a degree. I always liked working with little kids, so I picked early childhood. I was kinda scared cause I knew I really didn't know how to study. I kinda just got by in high school, but I knew college would be harder.*

Harold exclaimed:

> *I was psyched about starting college! I really hated high school. I got picked on, and I was sort of a loner, so it was nice to go somewhere where no one knew me, but my heart just stopped every time I thought of taking math. No biggie.*

Adam elaborated on some of his difficulties in high school and how such difficulties led him to college:

> *I got into some trouble senior year. I was out drinking with my friends, and I got a DUI. I got into trouble at school, but this was the first time I got into trouble with the law. My mom said I needed to go to school, or she was gonna kick me outa the house. That's the only reason I went; I hated school.*

Adam also shared some additional difficult life circumstances:

> *My dad died during my senior year from a heart attack; it was real sudden. I was just a mess; I couldn't focus on anything..*

Rosemary was enthusiastic about a change:

> *I had a tough time after high school. I had a lot of anxiety and depression and I drank too much. I was in and out of work, and I had some bad relationships. I was also on and off medication. My friend convinced me to go back to school and told me I should take some classes at Arnold, and she even helped me with all the paperwork. I was excited about going back to school. I thought I was finally getting my life together..*

Arnold Community College

Emily, Joyce, Audrey, Deb, and Otis all experienced anxiety as they entered Arnold Community College (ACC). Emily reflected:

> *I was so scared about going to Arnold [Community College]. I have social anxiety disorder, and it's hard for me to go to new places and talk to people. I hated school, and I had no intention of ever going to school again, but after high school, I just kept taking jobs, and I would get laid off cause they were making cutbacks. In my last job, I was a waitress at this bar, and this guy used to come in every night and say nasty stuff and grab me. I used to go home and cry. My mom had a long talk with me and convinced me that I needed to go back to school if I wanted a better life.*

Joyce shared her concerns as she entered ACC.

> *My husband wasn't too happy about me going to school. He didn't think I would have time to work and go to school, full-time, and be there for our little boy. That's why I took this class online. I figured if I was spending less time in class and going to class, I would have more time to study and be with mt family. But I knew; I wasn't a good student. How was I gonna have time to do everything?*

Deb knew that college would be different from high school:

> *I didn't really study in high school, but I passed everything. I knew it would be harder in college, and that scared me. I was really scared about math, but I was scared about all my classes.*

Otis spent thirty years working construction, but he said his body simply could not handle that kind of labor-intensive work. He wanted more of a professional career where he could better provide for his family. However, the upcoming transition weighed on his nerves:

> *I was having anxiety attacks the entire week before school started. The morning of my first class I couldn't stop throwing up. It was so many things. I couldn't believe I was going back to school after thirty years. I was going to sit in a classroom with all those kids. On top of that, I had to take math! Everyone was going to think I was just some stupid old guy.*

Audrey was anxious about starting school; however, she believed nothing could be worse than her life prior to community college.

> *After I dropped out of high school, I had a huge fight with my mom. I ran away, got in with this guy, and he turned out to be really abusive. One time he beat me really bad. I knew if kept this up, I would die, so I went back to my mom. We got a lot of therapy, and I decided to go back to school. I had to take an online class because I had a lot of anxiety. I couldn't be in large groups.*

Like some of the participants at FCC, Audrey prepared for her first developmental math course.

> *My aunt's a math teacher, and she said I should get the course materials, like the book, and start working on some stuff. This way I could be ahead of the class.*

With the help from her mother and aunt, Audrey got her math textbook along with materials online (e.g., worksheets and practice problems) that focused on the initial material in introduction to algebra.

> *I would just practice problems with fractions and signed numbers. My aunt would help me, and I found some great videos online too. It just helped relax me a little because I knew what was coming. The more I practiced, the more it came back to me. It was stuff I did in elementary and middle school, so it was just brushing up.*

Blair Community College

Mike was beginning college as a part-time student but working as a bank teller. He wanted a career where he had a chance to advance and could create a better financial situation for his family; however, this meant confronting his fear of returning to school:

> *Anytime I would think about going back to school, my heart would race. The week before school started, I kept thinking "I can't do this; I can't do this", especially about taking math again. I was willing to do anything my professors*

wanted me to do. I was willing to work as hard as I needed to. I just didn't think
I could do it.

Denise was recovering from a failed marriage, and she was ready to further her education to enrich her life.

I got married when I was eighteen, and I was miserable for eight years. I
sometimes had to work two jobs to support my ex-husband who was always out
of work and expected me to support him. Now it was time for me! However,
Denise had one fear: *I couldn't even imagine being back in math class again. I*
mean, how was I gonna pass math in college?

Joe shared:

Like I said, my parents made me go to school. I had to get financial aid, but they
said they would still help me out, and they said if I didn't go to school, they
would charge me rent, so I was like "whatever". I really didn't wanna go 'cause
I hated school.

Jerry shared that he and his wife were planning to start a family, and his wife urged him to return to school so that he could be a better financial provider, but he was not sure if going back to school was the right maneuver.

I was a horrible student; I couldn't wait to get outa school. I didn't wanna sit in
a classroom, do homework, take tests, and oh yeah, take math again!

Walsh Community College

Dina shared:

I was newly single, and I was ready to get my degree in art and make a better
life for myself and my son. I was more confident than I was back in high school,
but I was so scared about going back to school. Were all these young kids going
to think I was the weird old lady?

Dina recalled her anxiety, which stemmed from feeling lost and confused in math. Therefore, she looked for math preparatory programs to groom her for algebra 1. Unfortunately, all were online programs, and Dina wanted to work with a math professional in person. She talked with the administrators in Walsh Community College's (WCC's) math tutorial center, and they arranged for her to work with a student-tutor the week before classes. Dina elaborated:

It was great. They set me up with some videos and some worksheets that focused
on the stuff I would be doing in algebra 1. I also reviewed a lot of arithmetic, like
fractions and decimals. The best part was that I worked with this guy; he was a

student at Walsh [Community College], and he helped me so much. I practiced for a few hours each day. I was still scared about starting school, but I felt so much more prepared and so much relaxed.

Ron reflected,

I went through hell after high school, and I learned I needed an education. So, I was willing to do whatever I had to do. I was nervous coming to Walsh [Community College] for the first time, but anywhere was better than where I had been.

Tom shared:

I was excited about going to school. I wanted to be a pilot as long as I could remember. I just didn't understand why I had to take so may other classes, especially math.

Tara recalled:

I was so scared about going back to school and taking math, but I needed to go back to school. I just couldn't keep a job with just high school education. It was always the same thing. I would get a job and they would just keep cutting my hours. I think I went through eight or nine jobs after high school. I wanted a better life, but was college going to be like high school? I was so scared of sitting in a math class, feeling lost, and getting laughed at.

The Developmental Math Sequences and Modalities

Again, developmental math is content covered, or reviewed, in higher education that is also covered in the K-12 curriculum. Below are the developmental math sequences that are taught at community college as well as formats in which they are offered. For a full description of the content in each stand-alone developmental math, see Appendices B through E.

At FCC, the developmental math sequence consists of pre-algebra, elementary algebra, and intermediate algebra. All are half-semester courses, and all three courses are offered in the traditional lecture style, the emporium model (chapter 2), and online.

At Blair Community Collge (BCC), the developmental math sequence consists of basic algebra and intermediate algebra, which are both full-semester courses. Both courses are offered in the emporium model and the online format. BCC does not offer face-to-face formats for these courses.

ACC offers an arithmetic refresher course for students who do not test into introduction to algebra. This is a one-week intensive course that meets the week prior to the regular semester. Classes meet for three hours per day for four consecutive days. The content in this course focuses on operations with fractions, decimals, ratios and proportions, percentages, and introduction to integers. This is a lecture-based arithmetic refresher course; however, instructors tend to employ a lot of drill and practice as well as group work during class. Students take an exit exam at the end of week, and if they achieve at least 80%, they can enroll in introduction to algebra.

After the one-week arithmetic course, ACC has a three-course developmental math sequence consisting of introduction to algebra, elementary algebra, and intermediate algebra. All are half-semester courses and are offered in the traditional, emporium, and online modalities.

The developmental math sequence at Drummond Commmunity College (DCC) consists of two full-semester courses: introduction to algebra and intermediate algebra. These courses are offered in the lecture-based format and the online format. DCC also offers the Statway and Quantway pathway in two semesters. The quantitative literacy course is considered Quantway 1. This stand-alone full-semester course prepares students for quantitative reasoning, which is the college-level portion of Quantway. Both FCC and ACC also offer the Quantway and Statway pathways; however, these courses are in a corequisite modality, and this is explained in chapter 8.

At WCC, the developmental math sequence consists of algebra 1, algebra 2, and algebra 3. All are half-semester courses offered in the traditional, emporium, and online modalities. Students who enroll in algebra must also register for a mandatory SI session. The SI session meets once a week for fifty minutes and is instructed by an SI leader she attends classes alongside the students.

As for the course modality, Laura, Dan, Mike, Joe, Tom, Jerry, and Denise attempted their first math course in the emporium model. Joyce and Audrey registered for their respective courses in the online format, and the rest of the participants took their courses in a traditional face-to-face modality.

Starting Off on the Right, Wrong, and Indifferent Foot

Attempting the dreaded subject of mathematics in a new setting was certainly a major undertaking for these students. These students faced an arduous and uphill climb toward their goal of completing college-level math. A good start to the endeavor was imperative. Therefore, I inquired about the participants' initial experiences in their first developmental math course.

The Good

Several students reported positive experiences early in their math endeavor. At FCC, Jessica, Larry, and Cindy enrolled in the face-to-face modality for pre-algebra. Jessica shared:

> *I couldn't believe it; I was actually understanding what was going in math! I was doing fractions and decimals and the signed numbers. I got a 97 [%] on my first test. My teacher was awesome too. She just went really slow, and he explained everything so clearly. I think going to that head start helped me a lot 'cause we were covering a lot of the same stuff, and that [head start] really did get me a head start on the class. Another thing is my attitude was different. I decided it was college, and I need to take stuff seriously, like paying attention in class, taking notes, and doing homework.*

Larry had a high anxiety level regarding college and attempting math, but he was pleasantly surprised at the start of his college career.

> *I gotta admit; doing that head start program really helped. For the first time in my life, I was actually ahead of what we were doing [in math] and not behind. I even understood what was going on in class. Still, I was taking notes and listening to the teacher and doing my homework. Oh yeah, and I got an A on my first test.*

Larry also began to conquer some of his other fears:

> *So, I think it was the fourth or fifth class, we started getting into the signed numbers, and I got a little confused. I knew I needed to ask a question, but I was scared, but I raised my hand and I asked [the question]. You know what the teacher did? She smiled and said, "That's a great question", and she answered it really well, and I actually understood. You know what else? She was a great teacher too. She wasn't scary; she was so nice to us and made things make sense. No one laughed at me in class. It was so different than high school.*

Cindy and Todd were pleasantly surprised with the content of the course.

> *I didn't expect it to be so easy,* Cindy said. *The class started with what we were doing in that head start class and that was pretty much a review. I got a 90 [%] on my first test.*

Todd shared:

> *We started off doing stuff from like fifth or sixth grade. I was like, "I got this". I think I got a B on the first test. I just thought it was gonna be a lot harder, and all of a sudden the first day, my professor started doing fractions. A few of us looked at each other, like are we in the right class? I mean, I was rusty on some stuff, but as soon as the teacher explained something, I got it.*

At DCC, Harold and Rosemary were excited about their starts. Harold had a positive start in intermediate algebra:

> *The first day of class, my teacher made it clear that there was no messing around. He made it clear that if we didn't take this class seriously, we weren't going to pass. I mean; I was ready to take school seriously, but this guy made sure I did. Let me tell you; my professor just made class so interesting. Yes, he actually made math interesting! So, I was paying attention in class, taking it seriously, getting tutoring when I needed it, and I was even learning stuff.*

Rosemary felt that her educational endeavor was a breath of fresh air.

> *It was great! I had an amazing teacher who seemed excited about teaching math. I was going to class and participating. The math was hard, but I was there; I was trying; I was asking lots of questions. It felt so good to be doing something with my life.*

Taking an online class was new to Audrey, but she felt a connection to her classmates.

> *My professor set up these online study groups. These were people we could work with and go over problems with. We decided to Snapchat, and we all realized we had a lot in common. Have you ever met a group of people where you just clicked? We all hated math and were scared about this course. We all promised that we would get each other through the class. It felt so good.*

Deb, from ACC, conveyed that a one-on-one meeting with her professor eased her anxiety and allowed her to start her math journey on a positive note.

> *The first day of class, my professor told me that it was a requirement that we met with her outside of class in her office. It was basically a getting to know you. Again, I was serious about college, because I knew it would be harder, so I signed up right away. The meeting with her went so well. She understood how nervous I was about math. She encouraged me to take notes in class, and she told me about the college's tutoring center, and she said to come see her for office hours if I was confused. It was the first time a teacher reached out to me like that. I was actually keeping up with the class.*

Otis had an uplifting experience in the arithmetic refresher at ACC.

> *It was great. There were only seven of us in the class, so it wasn't that intimidating. The other thing was there were a few other people in their thirties and forties, so I didn't feel too out of place.*

While Otis entered the class with a great deal of anxiety, the professor and the dynamics of the class put him at ease.

> *I had such a great teacher. Her name was Mrs. Gallagher. She was so patient, and I honestly think she explained things better than anyone ever had to me. She was so calm, and she showed us how to go step-by-step and really think through problems. She also worked with us so much individually, and that helped so much. She also used a lot of real-life applications for fractions and decimals, which no one ever did, so it made sense. Again, my class was great. We worked together and helped each other out so much. I didn't feel so weird about going back to school as an old guy. If it wasn't for the support of the class and that great teacher giving me so much individual attention, I wouldn't have passed.*

Otis felt another positive volt when he passed his arithmetic refresher exit exam.

> *I was surprised, but I was happy; I got an 82%, so I just made it. It was scary to take a test for the first time in thirty years. My hand was shaking so much that I had trouble writing. There were some things I still didn't get too well, like those signed numbers, but I guess I did well enough. It made me feel better about the real math course the next week.*

Denise was surprised when she started her basic algebra class in BCC's emporium model:

> *At first, I was a little freaked out that I was going to have to do math on a computer. On the first day, I was ready to walk out.*

However, as time progressed, she began to feel that the dynamics of the emporium model met her needs.

> *I liked how I could get one-on-one help when I needed it. I didn't have to be embarrassed by asking questions. I also liked that I got so much practice in class. You know I remember in the past, something made sense when the teacher did it, but it made no sense when I did it on my own. The videos to help me understand the math were good too. The teacher who ran the class was the one who did the videos. But yeah, I felt comfortable going to class, and I was learning stuff.*

At WCC, the start of the first semester brought a sense of relief to Dina.

> *I couldn't believe it. My professor was explaining something, and I was getting it. I don't think that ever happened to me in math.*

Like other participants, Dina credited prior preparation to her smooth transition.

It helped that I spent so much time with that wonderful tutor, the week before, and practicing on my own and getting a head start. I think it helped that I was in a relaxed atmosphere when I was practicing too. If I had waited till school started, I think I would have been very overwhelmed.

Dina did, however, experience one source of frustration prior to school starting.

My academic adviser kept trying to get me to register for that math class in a computer lab [emporium model]. He kept telling me I could work at my own pace and get through the class quicker, but I had to keep telling him that I need math instruction from a live person. I had to pretty much talk him into it. Doesn't he get that I understand how I learn best?

For Ron, a positive attitude toward his studies made for a good start to WCC as well as a noticeable change in his math experience.

I went into college with the attitude that I was going to make this work. It's amazing how much more math makes sense when you go to class, pay attention, ask questions, and take notes.

The Bad

The math endeavor did not start well for some of the students. Dan and Joyce immediately struggled with time management. Dan was also struggling from a lack of sleep.

My wife works full-time, so we agreed that I would get up with our baby at night. Guess what? He got up a lot. I was just beat all the time, and it didn't help that my math class was at eight in the morning. I was so tired; I started missing classes.

Dan also discussed time constraints.

Taking care of a baby takes so much time, and I was taking four classes and working part-time. I didn't have time to do homework. The semester was slipping away from me already.

Learning math in an online environment was a challenge for Joyce who said:

I was getting some stuff but not everything. I got the rules for the signs [signed numbers], but I had trouble with the harder problems. My professor posted videos, and we had these discussion groups, but when I had a question, I just wasn't getting a direct answer.

Joyce was also struggling with time management issues:

> *I was at home, and everybody needed something. My son needed help with his*
> *homework; my husband expected me to do laundry, go grocery shopping, and*
> *drive my son everywhere. It was the first week, and I was falling behind. I was*
> *crying myself to sleep at night, because I wanted this [school] to work so bad.*

Emily became overwhelmed with the content in the arithmetic refresher very quickly:

> *I felt so stupid, and I was so totally lost in that class. It was just moving so fast.*
> *I didn't get fractions, and we had to move on. I didn't get decimals either.*
> *Everyone else seemed to be understanding it. It was so embarrassing, and I was*
> *even struggling with basic multiplication and division a little bit. I went home*
> *after the second day and just cried. I knew there was no way I was gonna pass*
> *that test [arithmetic refresher exit exam]. I couldn't believe this was happening.*
> *I mean, I was so scared to go back to school, and I was already lost. It was like*
> *my worst nightmare was already coming true.*

Emily touched on some other issues:

> *Like I said before, I have social anxiety disorder. Other kids were already*
> *starting to talk to each other and work together. It was so hard for me to talk to*
> *people, so I was always by myself.*

After some encouragement from her mother, Emily decided to speak to a school counselor about her struggles. Emily met with an academic counselor and a math department faculty member.

> *They were both so nice. Mr. Jones [the math professor] said it might be better if I*
> *started in the Adult Basic Education (ABE) program to work on some of my*
> *math and reading skills.*

The state-funded ABE program is free to the students. It is not part of ACC, and students do not receive any kind of credit or financial aid for attending; however, the two organizations work closely to help students.

> *I decided to go to the other class [ABE]. I still took another class at Arnold. I*
> *took a developmental English class there, but I figured this was the best way to*
> *help me with math, said Emily.*

Tara was in for a rude surprise when she arrived at her math class on her first day.

> *I got to my first math class, and my name wasn't on the teacher's list, so I had*
> *to leave and go the Registrar's office. There was a problem with financial aid, so*

I missed my first class. I was so scared about falling behind, and I was already behind.

Regarding her initial experience in algebra 1, Tara's experience paralleled Emily's.

I got to class, and I was just lost. They were doing signed numbers, and I didn't understand anything at all. I just sat there, and I was trying so hard not to cry. I was scared to go to school for three years because of feeling helplessly lost, and it was happening right away.

Otis had a positive experience in the arithmetic refresher at ACC. Unfortunately, a change in scenery marred his undertaking of introduction to algebra, the course following the arithmetic refresher. Otis recalled:

That [introduction to] algebra class just felt different from the start. It was about thirty students, and they were mostly eighteen or nineteen. I had to be the oldest one there. It felt like a high school class, and they were acting like they were in high school, laughing and carrying on. My teacher, Mr. James, had to yell at them in class to stop talking and quiet down. Mr. James just wasn't as nice as Mrs. Gallagher [the arithmetic teacher]. He had our names on a sheet of paper, and he would just randomly look at it and call on us. If we didn't get it right, he just smirked. The good thing was he started reviewing the stuff we did in the arithmetic class, so at least I wasn't lost at first.

Otis also reflected on an ill-timed and embarrassing incident that added to his discomfort in that math class:

Mr. James would have students go to the board, like two or three at a time, to write the answers to math problems every day. I think it was the second class; it was my turn. So, I'm up at the board, and Mr. James is sitting in one of the seats watching us. All of the sudden this young guy, I guess he worked for the department, came to deliver a message to the teacher. Guess what? He saw me at the board, and he thought I was the teacher. Of course, he did. I was an old guy standing at the board. I didn't know what he was talking about, but I guess Mr. James did, and he came up and talked to the kid. The other kids in class started laughing once they realized what happened. I was so embarrassed. I was already really embarrassed about being the old fart going back to school, and now this. Even Mr. James laughed. I left that day almost in tears, and I never wanted to go back to school again.

Kathryn, who missed the first three classes and was never able to catch up in the course, said:

I made up some lame excuse about a family emergency to my professor, but I just didn't feel like going to school. I was eighteen and enjoying being out of high school. I was just so sick of school. I would stay out late with my friends after work, so I just didn't feel like going to school. When I started going to class, I had no idea what was going on. I can't even tell you what we were doing.

Like Kathryn, Robert and Adam began their first semester with poor attendance; however, this stemmed from more personal reasons. Robert reflected:

I was dealing with depression really bad. It was so hard for me to get out of bed. If I was even going to class, I was coming in late. The whole class was a blur. I bombed my first test really badly. I even got special accommodations 'cause I have a learning disability, so I got extra time. It made no difference.

Adam shared his struggles:

I was still torn up about my dad dying, so I just didn't care about school, especially math.

I went [to class] the first day, and I was like, "enough with this". We were doing math I did back in middle school; I met some friends, and we started hanging out in the student lounge all day. It was kinda cool. We would skip class, and nobody gave us a detention.

Mike began BCC with a positive attitude but with a great deal of anxiety. However, he was in for an unwelcome surprise on his first day of class. Mike recalled,

I go to my math class, and it's in this computer lab [emporium model]. I thought I was in the wrong room. I find out the whole class is done on computers.

Unfortunately, the events got worse from there.

So, we have to register for the MyLab Math, and I was so lost. All these young kids around me are signing up with such ease. I felt like such an idiot. My professor had to basically hold my hand the whole time to get me set up.

Finally, Mike was not happy with the pedagogical approach to the emporium model.

I have to watch a video and teach myself. Are you kidding me? I'm horrible at math; I need a person to explain this stuff to me. So not only am I having trouble with this MyLab Math, but there is no way I can learn from a freakin' video! The whole class and what we were supposed to do was just so confusing.

Mike's frustration continued:

> *I go to the second class. My username and password [for the MyLab Math software] wouldn't work, and my teacher had to spend time helping me. I was already lost and confused, and I hadn't even started doing math. How could Blair not offer a math with regular instruction? What about older people like me that need that type of instruction? Who decides these things?*

Like Mike, Jerry experienced some shock with the emporium model.

> *So, all of us are in this class because we suck at math, and we're supposed to teach ourselves? Please explain that to me,* exclaimed Jerry.

However, Jerry's struggles paralleled more with Emily's in that he was adrift from the first day.

> *I had to have either the teacher or one of the tutors sit next to me the whole time in class to do math problems 'cause I wasn't getting anything. We started off reviewing fractions, I think, and I didn't remember anything. We could use the calculator, but that wasn't helping too much. I remember I had to do problems like 4 times what is 36? The calculator doesn't tell you what the "what" is. It would take me forever just to do one problem with fractions. I could tell the tutor and teacher were getting angry 'cause I should know it.*

Tom (WCC) had trouble in the emporium model as well, but his was self-inflicted.

> *I kept getting into trouble for talking in class. The math was just so boring. All we did was work on problems in class and I already did all that in high school. The professor said he would kick me out of class, so I figured I better stop.*

The Indifferent

The remaining participants, Laura, Andrea, and Joe do not possess negative or positive memories from their introductory time in developmental math. In fact, all three felt that community college was simply a continuation of high school. Laura expounded:

> *Taking math in the lab [emporium model] was a little weird, but really it felt more like high school. I went to class; I went to work; I went home. I mean we were doing a lot of the stuff [math content] that we did in like seventh or eighth grade. It was sorta annoying that I had to do all that again, but I liked that I could work at my own pace and not have to listen to some teacher go on and on.*

Andrea's initial experience in community college paralleled Laura's. Andrea reflected:

> *I was nervous at first about going to college and new classes and new people, but it pretty much felt like I was back in high school. I met new people pretty fast. We went to lunch together; we started talking on Facebook.*

Andrea recalled her preliminary experience in developmental math.

> *I guess it was kinda boring; I mean, we were doing signed numbers, and I did all that back in middle school.*

Andrea and Laura also shared the results of their first developmental math exam.

> *I think I got about an 85, and I didn't study*, said Andrea.

> *I remember that I had forgotten we had a math test that day, and I still got a B,* recalled Laura.

Joe shared his initial reaction to developmental math.

> *I guess I was a little surprised that we were doing math on computers, but we did that in high school sometimes too, so that was all right. We started off with some easy stuff like signed numbers but then when it got to fractions, it was hard. I didn't really care. It felt like I was in the thirteenth grade, and hey, I passed high school. I just come and do the work, so I can pass this, right?*

Summary

The participants expressed mixed reactions to starting a new school and embarking on developmental math. Some met the challenge with anxiety, others with enthusiasm. For some students, the start to developmental math was a relief, and for others it was overwhelming. For some, it was simply mundane. With these ranges of emotions, the participants were about to embark on a course where very few succeed.

6

Navigating the First Developmental Math Course

The Results Are In

The following participants successfully passed their primary developmental math class in the first attempt: Jessica (grade of A), Larry (grade of A), Dina (grade of A), Ron (grade of A), Cindy (grade of B), Deb (grade of B), (grade of B), Laura (grade of C), Mike (grade of B), Harold (grade of A), Audrey (grade of B), Tom (grade of C), and Andrea (grade of C). The remaining participants were not successful in their first try. Rosemary (ACC), Emily (ACC), Tara (WCC), Dan (FCC), and Joyce (ACC) withdrew from their classes, whereas Todd (FCC), Kathryn (DCC), Robert (DCC), Adam (DCC), Jerry (BCC), Joe (BCC), and Denise (BCC) failed their classes. These students eventually successfully completed their primary developmental course. Later in this chapter, I will convey the common themes for success for them as well as the other participants. However, I will first examine the shared themes that attributed to some students' failure in their first attempt at developmental math.

Common Themes for Failure for Developmental Math

Twelve of the participants in this study failed their first developmental math class. Todd, Joyce, Kathryn, Emily, Adam, and Denise passed the course in the second attempt; however, Rosemary, Dan, Tara, Robert, Jerry, and Joe did not successfully complete the course till the third endeavor.

Apathy

A few of the participants disclosed that they possessed a lackadaisical and downright poor attitude toward their math classes most or all the time.

These students discovered that community college was different from high school. Joe explained:

> *I didn't want to be there, so I would go to class but leave early or just not work that hard. I'd just play around on my phone, and no one cared. I never did any homework. It wasn't like high school where you got yelled at if you didn't do your homework. In the lab, you could just go at your own pace, so I didn't think I was behind or anything.*

The emporium model allows students to work at their own pace; however, there are generally hard deadlines for when students must complete certain lessons or modules. This was the case at BCC, and students had to complete a pencil-and-paper exam. This took Joe by surprise, and he explained:

> *I took the first test, and I was lost. I was like, "I never did this". My teacher made me take the test even though I hadn't covered most of the stuff in the lab. I was angry. I blamed the teacher. I think I got like a 30 on that test. It didn't bother me too much. I figured I would still pass.*

Joe shared additional information on his behavior in class:

> *My teacher let us listen to music on our phones when we were doing math work, so I did. I would have the MyLab Math stuff on when the teacher walked by, but then I would switch to Facebook or something else if she wasn't looking.*

I asked Joe if he was ever reprimanded in class for not doing his work.

> *I think once she [his teacher] told me to get back to work, but no, most of the time she didn't care*

Kathryn began the semester by missing classes due to lack of interest:

> *I was kinda scared. I thought my professor might yell at me, but he didn't really care. He just gave me the syllabus and some other stuff. I was like, "Cool. I can miss class, and no one cares". It was great; I could stay out late and party with my friends, and just go to class if I felt like it. I think in the first few weeks, I missed more than half of my classes. I would show up and just fool around on my phone.*

I asked Kathryn the same question as Joe regarding if her professor addressed her behavior.

> *He called on me once, and I didn't know the answer. He pretty much ignored me after that.*

For other participants, their progress sagged as the semester progressed. Todd explained:

> *I got a B on my first test, and I didn't really study'cause I knew all that stuff. Math was always hard for me, but I didn't think it would be that easy, especially in college! So, I stopped going to class and quit doing homework. It was so hard to just sit in class with this guy [his professor] going on and on about stuff I already knew. The first time I skipped class, I thought someone was gonna call my house like they did in high school. No one did, so I skipped more classes.*

Denise began the semester in the emporium model by working extremely hard and keeping on top of her assignments, but as the material became more challenging, it taxed her bandwidth. Denise explained:

> *It just got to be too much, all my classes, my job. As the math got harder, I needed more time outside of class. You know I had to spend more time on my homework. I just got so tired because there wasn't enough time in the day. As the math got harder, I stopped doing homework and even started missing class. It was too much, and I needed a break, and there was no break.*

Denise mentioned that her professor saw her slacking and tried to intervene:

> *He would come around and talk with us once a week about how we were doing. He said he saw I was falling behind and even missing class. He asked if everything was OK. It made me want to do better, but I just didn't have the energy and time to keep up.*

External Issues and Time Management

For some of the students, personal problems and life issues were barriers to success in their attempt to complete their primary developmental math course. Robert clarified:

> *I just kept sinking deeper into depression. I was sleeping all the time; my wife left me; I lost my job. I would go to class sometimes, but I would be very late. I would take notes, but most of the time I had no idea what was going on. I remember some of my grades. I think I got a 12, a 20, and a 30.*

Like Robert, Rosemary struggled with emotional issues:

> *I went off my medication, and I started getting really anxious and depressed. I also have ADD [attention deficit disorder], and I just couldn't focus in class. I mean; I started off excited to be there, but it just got hard to focus. I kept asking the professor to repeat himself, and I could I tell he was getting upset. The other kids in class started laughing. I just couldn't help it. Even if something made sense to me in class, I would go home and try it, and it just didn't make sense.*

In hindsight, Rosemary was able to pinpoint a specific learning difficulty.

> *I was so messed up; I couldn't get basic stuff like rounding to the nearest tenth of a percent, and I couldn't put two things together like when we had to solve equations. I would get real frustrated.*

Rosemary discussed her attendance issues in class.

> *I got in a bad relationship, and that made my depression worse, so I was sleeping more and missing class.*

Adam shared his continual struggle with grief:

> *I was mostly cutting class 'cause I didn't care. I would show up to class every so often. My professor never seemed upset that I missed class, so I didn't think anything was wrong. It was hard for me get to class; I just couldn't concentrate. In class, I would just look at my phone. I missed my dad so much, but I figured I had to keep going to school.*

Adam discussed his exams:

> *I was bombing my tests. It didn't make me feel better. I really don't know if I thought was gonna pass the class or not. I was just in such a funk.*

Adam described his level of despair.

> *I thought about killing myself every day. I missed my dad so much, and I saw no purpose in life. Yes, I was going to school, but I was a bad student, so what was the point in anything?*

Dan and Joyce continued to contend with home issues that hindered their schoolwork.
   Dan explained:

> *I wanted to do well; I really did, but I just couldn't do it. My son woke up sometimes two or three times a night. Sometimes he would wake up and be awake for hours. It was 6:00 a.m. before I got him back to sleep, and my wife works full-time, and she wouldn't get up with him if she had to work the next day. I was just so tired. I was missing class so much.*

Joyce continued to struggle with being a full-time student with a part-time job and a full-time mother and wife.

> *I had no time to study or do homework. I was either working at my job or taking care of my son. Honestly, I would try to study and do work at home, and my son or my husband would always need something. I fell behind on my*

homework, and I didn't have time to study. I didn't do well on my exams because I couldn't find time to go over and understand the math.

Jerry's homelife also continued to spiral downward.

I lost my job, and my wife kicked me out. We had been having problems, so it was a long time coming. I had nowhere to go, so I was living in my car for a while. I was collecting unemployment, so that's how I was paying for gas and food, and I was taking showers on campus. It's hard to focus on school when you don't even have a place to live.

Drowning in the Progressiveness of Math

Several participants, who failed their first developmental math course, discussed the perils of falling behind in a math course and attempting content without the proper prerequisite skills. Todd began attending his pre-algebra class sporadically after the first exam. This following unit focused on linear equations and applications, and Todd began to struggle. He said:

I missed I think two or three classes, and we were doing these multistep equations with fractions and decimals, and I was so confused. Then, we started doing word problems, and I got some of it, but I was missing a lot of pieces. You can't do word problems with equations if you don't understand equations.

Being in an online environment plus not being able to study or practice math outside of class prevented Joyce from being able to master the necessary concepts. Joyce provided an example:

We were doing linear equations in class. I kinda got some things, but I was struggling with the steps I needed to do to solve the problem, but then we got to two- and three-step equations. Then, we started doing equations with fractions, and it was like I didn't understand enough about the earlier stuff to do what we were doing now. I knew if I just got my questions answered, like in a regular class, I could get it. My professor had us in these online groups, and it seemed like everyone was getting it but me. I would ask my professor a question, and he would email me, but it really didn't make sense; you know. I realized I needed someone to just verbally explain math to me. I need to hear it!

Joyce summarized:

I learned a couple of things. I need someone to explain math to me, and I need time at home to practice and do homework.

Rosemary found that struggling to keep up with the content had a snow-balling effect.

> *As soon as we started something new, I was behind in like two minutes. Like we started finding slopes and lines. I couldn't find the equation of the line because I was still having trouble solving for something [a variable]. It always felt like the rest of class was just moving on without me.*

Denise also struggled with lines and shared another example of how the lack of one prerequisite skill can hinder the ability to master subsequent concepts:

> *I didn't get slope; I didn't understand anything about slope. So, then we started doing all this other stuff with slope, you know like equation of the line and parallel and perpendicular lines. I was lost.*

Joe attempted to take his exams in the basic algebra emporium model even though he did not complete the required material. Joe revealed that he would take his study guides for the exams to BCC's tutorial services with the hope that the tutors could simply provide him with a crash course. Joe provided an instance of how this strategy failed:

> *I remember this one time I was trying to figure out how to factor. The tutor asked me a question about signs, and I had no idea. He asked me another question about the signs; I still had no idea. He finally said, "You don't know how to do signed numbers; do you?" I said, "I guess not". He looked at me like I was crazy, and he goes, "Well then, you can't do factoring".*

It is noteworthy that while operations with positive and negative numbers (signed numbers) can be computed using a four-function or scientific calculator, these devises cannot help students apply the procedures of signed numbers to topics such as factoring expressions.

Tara felt out of place and behind from the first minute in algebra 1, and the situation did not improve:

> *I'm not exaggerating. I didn't understand anything. I didn't get the signed numbers. I didn't even understand fractions or decimals. I couldn't even do basic math, like multiplying or dividing some numbers, without the calculator. My professor would call on me, and I wanted to crawl into a hole because I couldn't even give anything close to a right answer. It wasn't like high school where people laughed at me. I left after about the fifth or sixth class, and I never went back.*

Repeating the Same Behaviors, Expecting Different Results, and Deflecting Blame

The students shared some of the behaviors and patterns that led to their failures.

However, many of the participants mentioned that they repeated such behaviors and patterns but anticipated better results.

Todd, Kathryn, and Joe continued to engage in sporadic attendance while putting forth minimal to no effort to catch up on their coursework. Todd explained,

> *I just figured things would work out. I did well on the first test, so I thought that would be enough and would do OK on the rest of the exams.*

Todd felt a sense of entitlement due to absences:

> *I was ticked at my teacher. In high school, if we missed class, we were kinda off the hook for stuff we missed. Now, my teacher expects me to learn stuff even when I'm not there. What the heck!*

Kathryn said,

> *I don't know what I was thinking, to be honest with you. I was just having a good time. I was eighteen, and I never had that kind of freedom before.*

Joe continued his behavior after failing his first math class at BCC.

> *I guess I was a little surprised that I failed. I mean, in high school I always somehow passed my classes, so I thought it would be the same. I thought maybe with a different teacher I would pass next time, but yeah, I pretty much did the same thing the next semester too. I just thought my teacher wasn't being fair. I don't know what to tell you. In my mind, I just thought I would pass.*

Other students continued to attempt their education without addressing their external and personal issues. Rosemary explained:

> *I started the same math class [quantitative literacy] the next semester with a positive attitude. I changed medications too, so I thought that would help. I was excited; I was participating, but the same stuff started to happen all over again. I had trouble concentrating, and I was getting so confused in class. I was getting depressed and anxious, and I was missing class. Oh, and like before, I wasn't trying to get any help, like tutoring, when I was struggling. Even worse, I went back with that old boyfriend, and he was even worse than before. He started hitting me and stuff this time.*

Both times, Rosemary explained her situation to her teachers:

> *I told both professors about what was going on in my life. They were nice about it, but they were like, "You still have to do the work". Then, after a while, they wanted me to drop [withdraw] the class! I was mad. I thought they should cut me some slack. I felt like I was trying.*

Robert also continued his education while battling personal demons. Like the others, he blamed his teacher.

> *I told my teacher how I struggled with depression. He never really responded. All he [Robert's teacher] cared about was that I did the work. Anyway, I registered for the same class [quantitative literacy]. I guess I just thought things would be different, but it was the same stuff all over again. I couldn't get out of bed most days. When I was in class, it was a blur. I tried explaining my depression to my next teacher, and she recommended me to a school counselor, but that was it. I admit it; I thought I deserved leniency because I was depressed, and it wasn't my fault.*

Jerry was honest about another reason to stay in school:

> *Yeah, I admit it. The financial aid I was receiving helped me pay for food and gas and stuff. I had no job and no place to live. So, I registered for more classes. I know I failed most of my classes the previous semester, but I still was able to get financial aid.*

Unhappy with the emporium model, Dan decided to try the traditional lecture-based instruction at FCC.

> *I couldn't understand learning from a computer. At first, I liked the normal class [lecture-based] a little better. I liked having someone explain it to me, but I got lost pretty fast. It was just like the lab [emporium model]. I was just missing basic skills, like multiplying and dividing in my head or what is the lowest number that 6 and 10 both go into? It seemed like everyone else in the class could do that in their heads. I couldn't even do it with a pencil, paper, and a calculator.*

Tara attempted algebra 1 again but found herself in the same situation.

> *It was a different class with a different teacher, but it was the same thing. I was just lost. I thought missing the first day of class my first semester threw me off, but that wasn't it. I still couldn't understand anything.*

Even the SI classes could not help Tara.

> *We did some arithmetic in those classes but mostly we just reviewed what we had done in class, and I was just so lost. I was as lost in the SI class as I was in the regular class.*

Like Mike, Jerry experienced some shock with the emporium model.

> *So, all of us are in this class because we suck at math, and we're supposed to teach ourselves? Please explain that to me,* exclaimed Jerry.

However, Jerry's struggles paralleled more with Emily's in that he was adrift from the first day.

> *I had to have either the teacher or one of the tutors sit next to me the whole time in class to do math problems'cause I wasn't getting anything. We started off reviewing fractions, I think, and I didn't remember anything.*

Jerry elaborated on his difficulties:

> *We could use the calculator, but that wasn't helping too much. I remember I had to do problems like 4 times what is 36? The calculator doesn't tell you what the "what" is. It would take me forever just to do one problem with fractions. I could tell the tutor and teacher were getting angry'cause I should know it.*

Was Failure in Developmental Math an Anomaly or a Commonality?

For the students who failed developmental math in their initial attempts, their outcomes in other classes were mixed. Dan, Robert, Jerry, Adam, Rosemary, and Joe failed most or all their other classes. This was generally due to these individuals exhibiting the same behaviors, such as skipping classes, not taking notes, not completing homework, and not studying, as they did in their math classes. Other participants achieved success in other classes. The reasons for this varied.

> *I passed all my other classes,* said Todd. *I got an A in my introduction to hospitality because I was really into the course. It was what I wanted to do, so yeah, I put forth the effort.*

Kathryn had a similar experience in her early childhood class.

> *I was actually interested in learning about children, so I would actually read the book for class. I also enjoyed the group and working with my friends in class.*

Denise clarified that some courses are more salvageable when falling behind:

> *I fell behind in my history class, but I was able to catch up. OK. I didn't read our textbook. I would search the topic on the Internet and get a brief summary. It was enough to pass the tests, but you can't do that in math. You either get it, or you don't. There are no shortcuts. You need to spend time on math outside of class.*

Joyce took two other online courses, and she explained how it was easier to pass those courses as opposed to math:

> *In my introduction to human services, my professor would post a Power Point, and it was easy to understand. I also had to read chapters, and they were pretty easy. Same with English. We had to read and write papers, but I could find time to do that. With online math, it takes so much time to really try and learn what to do. That was the killer. It's the time it takes to actually learn the material. And then you have to practice.*

Tara succeeded in her other classes, during her first semester, and she discussed how this gave her confidence.

> *I did really well in my English, my Spanish, and my psychology classes. That's probably the only reason I went back to school the next semester. I started to realize that I wasn't stupid. I actually started to like school and even made some friends. I realized I could be a good student, and I was learning how to study, but I just didn't know how to pass math.*

Themes for Success in Developmental Math

In the next section, I will convey common themes and practices that led all twenty-five participants to succeed in their first community college math class. More specifically, these were strategies that students attributed to helping them learn new content, keep pace with the course, and ultimately pass the course. Although such strategies contributed to their success on exams, I will specifically address successful exam preparation in chapter 8.

Addressing Personal and Mental Health Issues and Time Management

Several students acknowledged that they needed to address barriers in their own lives before they could conquer the challenges of mathematics. Adam shared how it took hitting bottom to rectify his educational endeavor:

> *I just totally broke down. My mom admitted me to a psychiatric hospital, so I could get help with my depression. I spent some time there and kept seeing a counselor. My doctor also put me on medication. After a while, I just started seeing things clearly. I still missed my dad, but it was different. I wanted to do good in life. I wanted to do well for him. I knew it was gonna take hard work, especially in math, but I was ready.*

Rosemary addressed her issues as well, but it took a stern talk from her math professor during her third attempt at quantitative literacy:

> *I showed up to class one day, after missing a few classes, and I got really frustrated'cause I was getting so confused. He [Rosemary's professor] asked me to stay after class, and he just went off on me, but in a good way. He told me I was really smart, and I had a good attitude, but I need to get my life together and stop blaming everyone else. It really made me think. I was thirty years old, and I spent much of my adult life just screwing up and blaming other people.*

Rosemary decided to tackle her issues in a matter-of-fact manner:

> *I talked to my doctor about my depression and how I couldn't concentrate, and he adjusted my medication. I was ready to take school seriously and take some responsibility for the first time in my life. You know what? Taking my medication and dealing with my issues, I felt more focused in class, and it made such a difference.*

Both Joyce and Dan needed to have serious conversations with their respective spouses regarding more support for their education. Joyce elaborated as to how this was not an easy task.

> *It took having a few conversations. I finally had to say to my husband, "I want to do this [go to school] to create a better life for me and our son. This is something I really want. Do you have my back?" He finally understood that I needed time. We made an arrangement that I would get an hour to two hours most evenings simply to study and do homework.*

Joyce further assessed,

> *I think my major mistake in going back to school was just assuming I would find time in a crowded day to study and do homework. You just can't assume that. You have to make time in advance.*

Dan shared:

> *After basically flunking my first semester, I had a long talk with my wife about needing more time for studying and sleeping. I guess it came down to just talking with her about what was wrong instead of being angry at her and blaming her that I didn't have time. She agreed to get up with our son a little more at night, but more importantly, we set up time during the day when I could sleep and study. We got my in-laws to help out more, and we hired a babysitter. It still was really hard. Going to school when you have a baby is the hardest thing in the world, and I was still really tired, but man, what a difference it made when I had some set time to sleep and study. Sometimes you have to stop blaming everybody and just realize you need to be honest and ask for help.*

Jerry realized it was time to ask for help.

> *I was about to be academically dismissed, and I had no job and no permanent place to live. I found myself going through the garbage one night to get something to eat. I almost got beat up by some homeless guy for going into his territory, and I just stopped, and I couldn't believe this was me. I decided I need to get help, so I went to see a counselor at BCC. She referred me to a guy in social services who was able to help me. He helped me get a job and acquire some temporary housing. I guess I was kidding myself thinking I could go to school with what I had going on.*

Robert got academically dismissed from DCC due to poor grades. He described his situation at the time.

> *I was sleeping all the time. I even started drinking, and I was blaming everyone else. I thought the world owed me a damn favor.*

Robert sought treatment for his depression, alcoholism, and began counseling with his ex-wife.

> *I had to think long and hard about a lot of things, and I decided I wanted to make schoolwork. I really did want go into human services to help people like me.*

Larry (FCC) and Audrey (ACC) needed to confront personal demons before starting their college education.

> *After high school, I basically had a nervous breakdown. I got picked on so much; I had so many bad experiences, and I was just afraid of everything*, recalled Larry.

Larry's parents got him intensive psychotherapy.

> *It really helped. I got on medication, and I just started learning how to face the world without being so scared. Without those people who helped me, no way I could have even come on to the Flores campus*, explained Larry.

> *I just thought I was stupid in middle school and high school*, explained Audrey. *I figured out I wasn't stupid. I just couldn't concentrate on math because I had so much going on. That just made me human.*

Several participants realized that tackling the discipline of math, let alone going to college, not only takes extraordinary time management but also takes sacrifice to ensure that there is time. Joyce, for example, worked out a schedule with her husband so that she could study, but she also made some sacrifices.

I had to cut back on my course load, especially for the semester. I was taking math. So, when taking math, I took only two courses instead of four. I also cut back on some of my work with the church on weekends.

Denise spoke with her academic counselor regarding the problems she encountered as her first semesters progressed.

He [Denise's counselor] thought I was taking on too much. I was working full-time and taking a full course load.

Denise had a dilemma. She was using federal financial aid and thought she needed to carry a full load to qualify. Denise's counselor directed her to the financial aid office. Denise explained,

I was so relieved. I was still able to get some financial aid even if I took less than 12 credits. So, I only took six credit hours the next semester so I could focus on math.

Of course, it took some additional sacrifices as well. Denise elaborated,

Yeah, I had to cut back on going out dancing and drinking with the girls and some of my reality TV. It made such a difference. With more time, I had more energy and concentration to study more and do better in math.

Otis (ACC) explained how his wife helped him devote time to his studies:

The first weekend after school started, I lay down on the couch to watch football. My wife came over and shut the TV off. She told me that I needed to do a better job setting aside time to study. If I was gonna make this work, I had to give up some stuff I liked to study. You know something? She was right. If I didn't organize my time, no way would I have passed math.

Addressing Learning Issues, Prerequisite Skills, and Learning Styles

To be successful, some participants had to address barriers such as deficient basic skills and learning differences. As discussed in chapter 5, Emily struggled in the arithmetic refresher at ACC and was redirected to an adult basic education (ABE) class to further develop her fundamental math skills. Emily elaborated:

I am so grateful they recommended that class. Those people were amazing. The [ABE] class was run by two women, and sometimes they would teach the whole class; sometimes they would let us just work on problems. I was able to just focus on things like multiplying, dividing, fractions, and decimals. There was no rush; I could just take my time. It also helped that everyone else in the class was having trouble with the same stuff. The best part was I was starting to understand math for the first time in my life! I wasn't just memorizing stuff either; I was really understanding it.

Emily provided an example:

I finally understood greatest common factor and least common denominator. I mean, I finally got why 8 is the greatest common factor for 16 and 24, and I could even explain why to someone else! And it was more than that. I was starting to do problems like that in my head. I could never do that before. I was starting to see that do to algebra, you have to be able to do basic math in your head.

After Emily completed the ABE course, she shared a pleasant surprise.

Guess what? When I retook the placement test at Arnold Community College, I tested into introduction to algebra.

Like Emily, Tara struggled with basic math skills from the beginning. In her second attempt at algebra 1, her professor took an interest in her education. Tara elaborated,

One day Mr. Willis asked me to come to his office. He could tell I was really having trouble, and he asked me to take this basic arithmetic test. He said it would just help him understand what I needed and how he could help me better.

Tara explained how this was the first step to success.

Mr. Willis met with me later. He told that me that he thought I was smart and that I was a good student, but I really needed work with my basic math skills. I needed to work on multiplication, division and just basic stuff before I could do algebra. It was the first time a teacher reached out to me and showed faith in me, which was nice. Anyway, I talked it over with my parents. We met with a school [WCC] counselor, and he recommended, I take a math class through an adult education program.

Tara enrolled in an external and state-funded adult basic education class.

The people there were awesome. It was great just to learn basic math and not feel stupid. I actually spent two semesters in the program. I just needed the time. My skills were that bad, but it made such a difference when I started algebra 1 at WCC.

Tara explained how her teacher in the ABE class employed a constructivist approach to help her master her math facts and develop her number sense:

Every day, Jane [Tara's teacher] would write a number on the board, like 42. She would have us come up with all the ways to get 42. So, there is 6 times 7; there is 30 plus 12; there is even six times six plus 6. We had a lot of fun as a class and we supported each other. This one time, she put up 192, and I came up with 120 ÷ 2(3) + 12 and everybody clapped for me.

I asked Tara how she knew she was ready to attempt algebra 1 again at WCC.

Mr. Willis gave me a bunch of questions from arithmetic. There were questions on multiplying, dividing, fractions, decimals, and common factors. He said when I could do these problems on my own without the calculator, I was ready.

Jerry had to tackle personal issues before continuing his education. However, he had to address academic issues as well. Jerry also talked to a learning specialist.

I have a condition called "math dyslexia". It basically makes learning arithmetic very difficult, and that's why I can get so confused so easily. I got special services in high school, but I have no idea if I was diagnosed with this or if they just thought I was an idiot back then. Anyway, my counselor said they [learning specialists at BCC] would work together to help me with math.

Consequently, like Emily and Tara, Jerry attended an off-campus ABE class as well.

The class was great. I was able to just focus on basic math. Then, I worked with a learning specialist at BCC who helped me as well. She [the counselor] helped me organize my notes and stuff. That's really what I needed, more practice, better organization. When I left the class, I felt like a different person. I was ready for algebra! said Jerry.

Jerry's experience in the ABE courses paralleled Emily's and Tara's in that he was able to gain an in-depth understanding of basic math and develop number sense.

I couldn't believe it; I could add and subtract fractions and even explain what I was doing out loud.

Dan found himself struggling with basic arithmetic skills as well. Consequently, Dan spoke with his academic counselor about this, and his counselor conferred with a professor in the math department. The math professor gave Dan a basic arithmetic test to assess his strengths and weaknesses. The math professor and Dan's counselor worked with the people in FCC's tutorial center to construct an individualized program for Dan. During the summer, he spent time working on basic arithmetic skills such as basic math facts, fractions, and decimals at his own pace while receiving assistance from FCC's student tutors. Dan explained:

It was such a big help to just spend time on that stuff and get that extra help. After a while, it started coming back to me. It didn't turn me into a math whiz or anything, but I was so much more ready for pre-algebra,

Rosemary had to face learning issues as well.

I was diagnosed with adult ADHD, and I started taking medication.

Rosemary explained the difference when attempting math while receiving the proper treatment.

People don't understand. When I wasn't on my medication, I just couldn't focus; I was trying to understand stuff, but my mind was just spinning. You know those games shows where someone is in a tiny room and they are trying to grab all the flying money, but it's so hard? That was me in a math class. All this stuff is flying around so fast, and I am trying to grab it, but I just can't. Being able to focus just slows things down and makes such a difference.

Denise, Joe, Mike, and Dan had strong but varying feelings regarding the emporium model. Denise shared:

I always liked the lab, but now that I learned to organize my time better, it was great. I don't think I would have passed basic algebra if I wasn't in the lab class. It was so great to get that great individual attention when I had questions and get all that practice in class. It was so much less pressure that the regular [lecture-based] math classes I had taken.

Joe also asserted that the dynamics of the emporium model helped him pass basic algebra.

I like how I could spend more time on the stuff that was hard for me and less time on the stuff I got. I got the evaluating expressions and the [linear] equations pretty good, but I really had trouble with factoring. So, I got to spend more time on factoring. In a regular [lecture-based] class, we would have had to move on, but I got to spend the time I needed on factoring.

Joe also appreciated the flexibility of the math lab at BCC.

You could go in there anytime the lab is open to work on your math. Even if it wasn't my class time, I could just go in there and work on my math and get help from a tutor. When I was doing factoring and those fractions [rational expressions]. I was in there a whole lot.

As discussed in chapter 5, Mike struggled with the emporium model format, and it did not get easier:

> *After the third class, I knew this wasn't going to work out. I was having enough trouble with the computer software; how was I gonna learn the hardest subject for me?* questioned Mike.

Mike was especially frustrated that BCC did not offer lecture-based instruction; therefore, he started researching his options. He opted to take introduction to algebra at DCC, which is roughly twenty-five miles from BCC. Fortunately, the class at DCC started one week after BCC, so Mike did not lose too much time.

> *I was straightforward with my counselor at Blair [Community College]. I told him that I can't learn math this way, and they were gonna lose me as a student if they didn't let me take math the way I needed it. So, he worked it out that I could take the math class at Drummond,* said Mike.

He shared his experience at DCC.

> *It was like night and day. It made such a difference to have someone just explain stuff to you. No way I would have passed math in that computer lab.*

Dan needed to face homelife issues before undertaking pre-algebra again at FCC.
However, he knew the kind of learning environment he needed.

> *It's pretty simple. I don't know algebra. When I don't know something, I need someone to teach me. It needs to be a person, not a computer, not a video.*

Dan provided an example:

> *I had never done word problems in algebra before. You know "let x equal this and let x + 4 equal that?" I don't know; maybe I did them and blocked them out. Anyway, I learned how to do them'cause I had a good teacher explain it step-by-step clearly, and I was able to ask questions. You know what else helped? We were all [Dan's class] doing it together, so other students had questions too that helped me out. No way I learn that in the lab from a computer or a video.*

Like Dan and Mike, Joyce benefited from a face-to-face class.

> *It makes such a difference when you can ask questions while your teacher is going through a problem. In the online class, I would watch a video with let's say five steps to solve a problem. If I got lost in step three, I was done, and I couldn't ask questions. In class, I could ask my teacher when she was finished with step three, and then I could understand steps four and five.*

Audrey, however, asserted that she would not have survived a face-to-face class, and the online format was imperative for her starting school.

> *I was basically held prisoner by my boyfriend, and he would beat me. I was afraid to go out and be in crowds. My doctor said it was agoraphobia. So, this was the only way I could go to school.*

Quality of Life as a Motivator

In chapters 4 and 5, several participants noted the quality of life without higher education (e.g., unlikable jobs, low pay, and no chance for advancement) as an incentive for pursuing higher education. This was also a motivator for some students to persevere through their developmental math classes. Jessica (FCC) clarified:

> *I remember when we got to solving equations [in pre-algebra], I got really stuck, especially when we had to solve equations with fractions. I just wasn't getting it at all. I felt so hopeless, and I wanted to quit. Then I remembered what my life was like as a waitress at Chili's. I remember all those nasty customers, my mean boss, how my legs and back ached after those long shifts. If I didn't have those horrible memories, I would have quit.*

Emily's experience after high school paralleled Jessica's:

> *It was scary going back to school, and when I had problems in that arithmetic class, I wanted to quit school. My mom reminded me of that horrible waitressing job with those horrible people. I used to cry myself to sleep thinking that this was going to my life forever. Going to school, and even taking math, reminded me that I could have a better job and life.*

Audrey used a horrific relationship as a driving force.

> *Every time I struggled and got frustrated with math, I reminded myself of how Rick [Audrey's abusive ex-boyfriend] made me feel like I was nothing. But I was gonna show the world that I was something.*

Otis worked several outdoor jobs throughout his adult life. In addition to bad memories, his body reminded him that he needed a change:

> *I had bad knees and an aching back. I also remember feeling the sweltering heat, and I remember my skin hurting because of the bitter cold. Math is scary and frustrating, but at least it doesn't make your body hurt. I reminded myself of that when things got tough, and I wanted quit school*, shared Otis.

Denise found herself at a crossroads when she failed her first developmental math course. However, it was her life prior to enrolling in college that motivated her to adjust to a schedule and life so that she could devote more time to math. Denise explained:

> *I had a lot of crappy jobs, but the worst thing was my ex-husband never wanted me to go to school. I guess he was afraid of me being better than him. So, I was a miserable wife with a horrible job, and I was only in my twenties. When I failed my math class, I had visions of going back to that horrible life. I'm gonna get my degree so I can make something of my life.*

Kathryn and Joe decided to drop out of college after their first and second semesters, respectively. Kathryn explained:

> *Like I said, I was having such a good time going out with my friends and hanging out. I was eighteen, so my parents didn't give me a curfew anymore. I didn't need school, and I definitely didn't need math. So, I was able to expand my waitressing job to full-time.*

After some time, Kathryn began to tire of the life she wanted.

> *My parents made me pay rent because I wasn't going to school. I also bought a new car, and my parents said I was on my own for car payments and insurance. Working full-time as a waitress wasn't enough. It really started to depress me after a while; this was gonna be the rest of my life?*

Kathryn decided to search for different employment but found this difficult with merely a high school diploma.

> *There just wasn't anything out there. I finally got hired as an aide at a daycare center. The money was still pretty bad, but I liked the work better. It made me remember how I was going to major in early childhood and work with kids. I mean, I was working with kids, but I wanted more. I wanted to be in charge, maybe run my own day care center to my own preschool. So yeah, I decided to go back to school.*

This provided Kathryn with the incentive she needed to approach her education differently.

> *No more cutting classes; no more goofing off. I registered for introduction to algebra, and I was ready to work hard.*

Ron (WCC) explained the difficult times he faced after high school and how this set him in the right direction.

> *I hated high school, and I wanted to get as far from home as possible. I just wanted a different life. So, I moved away and got a job. You know what? I wasn't as easy as I thought. I lost my job, and I couldn't get another one. So, I*

got in with this bad crowd. We had this stupid idea to break into a store to get money. We got caught, and I had to spend time in jail. Let me tell you something. There is nothing as scary as being in jail. It's like time stands still. And this one day I must have looked at somebody the wrong way, and this guy beat me up really bad. Anyway, I got probation, and I moved home. I decided to get my life together and go to school. Math is scary, but nowhere as scary as being in jail, and anytime I got frustrated with math or school, I just remembered where a life with no education got me.

Joe was academically dismissed from BCC. After some soul-searching, he decided to join the army. Joe described his experience:

At first, it was terrible. I was a complete mess, but I had a drill sergeant who kicked my butt, and after a while, the army made a man outa me. I finally learned some discipline and to take things seriously. Actually, if you are gonna make it in the army, you don't have a choice.

Joe developed a skill in the army as well.

I started working in the kitchen and realized I loved to cook. I mean I would cook here and there before the army, but I never really thought about it as a career or anything. Anyway, I worked to become a culinary specialist. When I was in the army, I decided I would go back to school and get a culinary arts degree so that I could do even better in life. Maybe I can open my own restaurant someday.

Joe described how the army prepared him for the dreaded subject of math.

I became a different person. I had goals, and I believed in myself. In the past, if I started a math problem, like factoring, I would just think, "I'm never gonna get this". After the army, I was like, "So, how am I going to figure this out, because I will get it".

Surviving basic training helped Joe face math in another way.

Anytime I faced a really hard topic, I just remembered how I used to have to run three miles with full equipment in the heat. If I do that, I could find the equation of a line.

Making Connections

Several students noted the importance of establishing personal connections with others as a major part of their success in their initial developmental math course. This included bonding with classmates and a support network outside of class.

Unfortunately, some students experienced initial isolation and other social issues that caused anxiety. Earlier, I mentioned how Emily and Larry initially felt isolated in their start. Mike's initial impression of BCC was that of an aloof student population:

> *All the students walked around with earphones, listening to music. Either that or they were on their phones. I know I sound like an old guy, but I remember a time when people talked to each other. I would try talking to people, but they weren't interested. If this was college, I didn't like it,* recalled Mike.

Dina (WCC) also found frustration while attempting to connect with classmates.

> *I would try to get to know people in my class, and they just didn't respond. I would ask them about their majors or about the class, you know small talk, and they would give me yes or no answers and go back to looking at their phones or listening to their music. Going back to school was scary, and I really wanted to connect with people,* shared Dina.

While Otis had a positive experience in his arithmetic refresher, his start in the introduction to algebra course was blemished with him feeling out of place and an embarrassing incident that almost led him to terminate his education.

> *I felt so out of place in that class; the teacher was a smart aleck, and after that guy thought I was the teacher, and people laughed, I never wanted to come back.*

Otis remembered his positive experience in the arithmetic refresher, and he decided to visit Mrs. Gallagher, the instructor for that course, and discuss his concerns. After consulting her colleague, Professor Jackson, Mrs. Gallagher suggested that Otis transfer to Professor Jackson's evening class. The class was smaller and had more nontraditional students. Otis recalled:

> *Wow! What a difference! Professor Jackson was so much better than Mr. James. He was so much more friendly and less intimidating, and I loved the class. I got to know two other classmates who were about my age. We met outside of class to work on problems, and we talked on the phone too. We got each other through the class.*

Otis provided an example of classmates helping each other.

> *My friend Mark really helped me with polynomials. I was really getting confused with all the parenthesis and the signs, but he just worked with me and helped me with some things I was doing wrong.*

Like Otis, Mike transferred math classes in his first semester. Instead of day to evening, he went from BCC to DCC. However, Mike still found a connection with his classmates. Mike's professor suggested that students meet virtually to work together. Mike elaborated:

> *A few of us were older returning students, and we just clicked. We had the same fears about math. So, we would meet a couple times a week using Zoom. Most of us worked, and I wasn't even from Drummond, so that Zoom was a lifesaver. We had this system where, when we met, each of us would bring up a problem we were having, and the others would help. I wouldn't have understood dividing polynomials if it wasn't for those meetings.*

Mike conveyed how meeting together instilled a form of student responsibility:

> *One of the reasons I was a bad student in school [elementary and high school] was I rarely did homework. Since I knew I was going to meet with other students, there was pressure to get the work done. If I didn't, I was pretty useless.*

Mike also noted the irony of the situation:

> *I get it; I was running from technology, and I wound up using technology to help me pass, but it's different. I had a great teacher who explained things, and this [Zoom meetings] was a way for us to help each other with what we already learned.*

Harold (DCC) shared how connecting with classmates and working collaboratively alleviated math anxiety.

> *My professor would encourage us to work together in class. I wound up with a study buddy. Her name was Sarah, and we started working together outside of class too. I never knew this, but math is so much less intimidating when you attack it with someone instead of by yourself.*

Emily was fortunate to have a classmate organize a study group in her introduction to algebra class at ACC.

> *I was kinda nervous. I'm so shy, but I figured I would join. It was great. There were other people in the group that had bad experiences and were scared of math too. We met like twice a week and online and helped each other.*

Audrey (ACC) explained how her study group in her online introduction to algebra class became like family:

> *We got each other through the course. When one of us was having a hard time understanding something, the rest of us helped. I couldn't believe I was helping people understand math. Me! I never knew how much it helped to explain math to other people.*

Other participants found a support network outside of class. Jessica, Larry, and Cindy attended the head start program for pre-algebra at FCC. The head start program was part of FCC's math tutorial center where students can seek extra assistance. Larry clarified:

> *During the head start program, I met some of the tutors and the staff at the tutoring center. At the end of the week, they felt like family, so I would go there [the tutoring center] every day after class and do my homework. I would go there even if I wasn't having trouble. It was great to just be able to do my work, talk to them, and ask questions if I needed, and it did help because there were times I ran into problems doing my math homework, and there was someone there to help me.*

Cindy had a similar experience with FCC's tutoring center.

> *I actually met some people in head start, and even though we were in different classes, we agreed to meet in the tutoring center. I even met more people.*

Cindy explained how FCC's tutorial center assisted her both academically and socially in a new environment:

> *It's like I said; I'm really shy. It was scary to ask questions in class, so thank God I could get help there [tutorial center]. It was just so much easier to ask questions there and work with people. Also, it's hard for me to go to the cafeteria by myself or join a club or something. So, the tutoring center was a way for me to socialize, meet new people, and work on my math and get help. I even met some people and we started doing stuff outside of school.*

Dina found support outside of class as well. She joined WCC's student forum, a form of student government made up of non-traditional and part-time students.

> *I met so many people who were like me. They were older and were nervous about going back to school. Suddenly, I felt like I wasn't alone. I met two other ladies who were taking algebra 1 and we started working and studying together,* shared Dina.

During her first semester at WCC, Tara joined the international club. Although this club was not math related, she explained how this connection contributed to her success in math.

> *Even though I had to drop math, I started making friends and getting comfortable at WCC. For the first time in my life, I felt like I fit in somewhere. I met so many nice people in the international club and even some good friends. These people even encouraged me to keep going to school and face my math fears. Without them, I may have dropped out of school all together.*

Engaging and Effective Instructors

An overwhelming theme for success that emerged was faculty who were both engaged and effective in their pedagogical approaches. In chapter 5, Deb (ACC) shared that her introduction to an algebra professor required students to meet with her at the beginning of the term. Additionally, Deb's professor required one-to-one conferences in the middle and at the end of term as well:

> *I never had a math teacher who actually cared about how I was doing. Because she cared, it made me care about how I did.*

When Adam returned to DCC, he knew he wanted to be a psychology major. He decided to complete introduction to statistics, which is generally required for psychology majors. This meant he needed to complete Statway 1, the developmental component of Statway. Obviously, the course content differed from introduction to algebra, but the professor was vastly different as well.

> *I had Mrs. Johnson, and she really cared about how I was doing. She would collect homework and grade it and give it back, but get this, she would remember which problems we got wrong. After class, she would say, "Hey, Adam, do you understand what you did wrong on number 8?" I mean; this woman has probably, what, hundreds of students? And she remembers how I did on number 8? It made me want to do better.*

Adam referenced the previous professor he had for introduction to algebra:

> *I don't even think that guy knew my name.*

Upon returning to BCC, Joe enrolled in basic algebra, which was still in the emporium model format. Like Adam, he noted a difference in his professor.

> *Oh, it was like night and day. The lady I had the first two times didn't seem to care. This guy [Joe's new professor] would sit with each of us during class, at least a couple of times, to see how we were doing. And man, he would keep us on our toes. When he came around, and we were working on a problem, he would start asking us questions like "What are you gonna do now?" or "Why are you doing that?" He also got to know us too. He would ask how our other classes were going and how we liked BCC. I mean, I had a different attitude, but my professor made a difference.*

Harold discussed how his intermediate algebra made class engaging and even enticed him to continue to come to class.

First off, this guy just loved math. I don't think anyone ever got so excited over rational expressions ever! It made me want to learn and kept my attention. The guy was also a goofball. Out of nowhere he would toss a nerf ball at one of the students. He would even break out into song. I mean; he was a great teacher first and foremost, but he just made math fun.

In general, the students attribute their success to stellar teaching:

Professor Salzman just explained everything in detail from start to finish so thoroughly, said Larry.

I used to be scared to get answers wrong. Professor Jackson would actually get excited if we said the wrong answer in class. He would use it as a way to help us learn, recalled Otis

Mike reflected on his class at DCC:

Remember I told you about my eighth grade teacher who just said "I made it up" when I asked a simple question about graphing. I couldn't believe it. In my algebra class, we started doing graphing with lines, and that topic came up again. You know where you have an equation like $3x + 6y = 12$, and I'm like, "Oh no! I can't believe this is happening again". But you know what? Professor Holton actually explained that we can plug in any number we want for x and then solve for y, and vice versa, and he explained it so clearly.

Emily explained:

My teacher was so patient, so kind, and just explained everything so well. She got me to like math! Guess what? She got me thinking I wanted to become a math teacher. Yes, me!

Rosemary reflected:

You know why my teacher was so great? Because he just kept trying to find different ways to help us learn. When our class would struggle with a topic, he would just find another way to explain it to us.

Jessica shared:

I liked how Mrs. Kenney mixed the class up. She would lecture, and she would teach things really well, but then she would let us practice problems in class, and she would walk around and help us and give us feedback. I left class feeling like I understood stuff.

Robert praised his professor:

> *My professor explained things so well, but it wasn't just boring lecture. She interacted with us and would ask us lots of questions while she was teaching, and we would work on problems in class.*

Dina emphatically explained how her algebra 1 professor utilized the three p's to help guide her to success:

> *Professor Bordi used preparation, patience, and precision. First, I swear this guy must have practiced how to teach us each problem over and over. I swear he was the most prepared teacher I have had. It's like he knew what our questions were before we asked. Second, Professor Bordi was so patient with us. He didn't get frustrated when we didn't get it or had questions. Do you have any idea how much that helps people like me who were scared of math? Third, there was precision. This guy left no detail out when he explained how to do a problem. He was the most thorough teacher I ever had.*

Audrey shared effective pedagogy from her online class:

> *Mr. Dixon [Audrey's online professor] just did everything to make sure we understood the material. He made great videos that he posted where he ex-plained the section, but anytime we had a question, he would meet with us using Zoom. We could show him our work, and he would help us. You know something? Mr. Dixon made me want to be a better student. I mean, he tried so, so, so hard to help us, I wanted to try harder in his class.*

A few of the students discussed a technique that helped them acclimate very quickly to their math classes. Dina explained:

> *During the second class, Professor Bordi gave us a quiz on where everything was in the class. It basically had where our online homeworks were located, where the online notes were. It had questions from the syllabus. He even asked us questions like where his office was and where his office hours were.*

Tara's professor utilized this technique as well, and she explained how this helped.

> *People don't understand how hard it is to start college. You have all these classes with different professors, all these different requirements and it feels like there is stuff everywhere. I like how I was just forced to learn where everything was and what was on the syllabus right way.*

Ron added:

> *It just saves a lot of time when you know where to find your assignments, when things are due and when your professor's office hours are. Thanks to my math*

professor, I could use that time to actually do math. In my other classes, it was a few weeks into the semester, and I was still struggling and wasting time to find all that stuff.

Audrey explained the importance of the "where is everything" test tactic for an online class:

At first, I thought it was weird that we had to take a quiz on the syllabus and that basic stuff, but I'm glad we did. In an online class, it can feel really overwhelming at first. It was nice to understand the goals of the class and where to find everything.

Learning Mathematical Organization

Another common strategy for success was the participants' development of the mathematical organization in their first developmental math course. Deb elaborated:

I was so ready for my first test. It was on all the signed numbers and evaluating stuff, but I couldn't believe it; I only got a 65%. It was like, for the first time, I am understanding how to do math, and I know what I'm doing, but I get the wrong answers.

After establishing a good relationship, Deb decided to seek advice from her professor.

She [Deb's professor] said it was all about my organization. With evaluating expressions, I wasn't writing enough stuff down on paper, you know, the steps. She was right. I was doing the signed numbers in my head and on the calculator, but I wasn't writing it down. My teacher said not writing enough stuff down can make you forget or get mixed up.

Deb began to follow her professor's advice and witnessed a difference.

It was unbelievable. I aced my next two exams. Even if I could do the problem on the calculator or in my head, I had to remind myself to write all the steps down. This also helps me go back and check my work, so I don't make stupid mistakes.

Emily visited her professor's office hour before their first exam. She recalled:

We were doing those signed number problems with order of operations with fractions, and I just kept getting stuff wrong. My professor looked at my homework and said, "Where's your work?" I said, "I know the rules of signed numbers; I'm just getting it wrong when there are so many steps." She said, "No, you need to show all of your steps. Write down each step to show how you got from A to B."

Emily took her professor's advice.

> *It worked! Yeah, it's more work to write all the steps down, but it's totally worth it to get the right answer.*

Both Denise and Joe shared how showing organized work was imperative for their understanding of factoring expressions.

> *You know how you have to do all that guess and check, or trial and error?* asked Denise. *Well, I was getting frustrated'cause I was forgetting what numbers and signs I tried. I would just erase and try again. My teacher, Mr. Shore, told me to write down everything I tried that didn't work. I started getting it right!* (see Figure 6.1)

Joe explained:

> *For me it was just bad habits. When I did factoring before in high school, and when I took this class the first two times, that's how we learned it, to just erase and try again. When my professor told me to keep track of everything, it was like a light bulb went on!*

For other students, showing their work was not the issue; it was showing it in an organized fashion. Dan clarified:

> *My problem was I was writing my work all over the place. One day I was doing a problem; I think it was solving equations. My teacher came around, and he said, "Can you follow that [Dan's work to solve an equation]?" I tried to explain it, but I was like, no. So he showed me how to write out my steps going down [vertical organization] till I got to the answer. Man, do I appreciate him doing that!*

Otis had to make a similar correction:

> *You wanna hear something stupid? You know how on a worksheet or handout, you have a tiny amount of space to work out your problem? Well, I was trying to cram all my work into that space. It's just what I always did, and I didn't*

Factor: $2x^2 - 3x - 5$
Try $(2x + 1)(x - 5)$
Does not Work
Try $(2x + 5)(x - 1)$
Does not Work
Try $(2x - 5)(x + 1)$
Worked!

FIGURE 6.1
Organizing the Trial and Error for Factoring Trinomials

realize how it was messing me up. It was actually Mrs. Gallagher [Otis's arithmetic professor] who kept after me about using separate paper and writing all my steps really clearly.

Larry has a lot of gratitude for the staff in the FCC tutorial center:

It was during the head start program. One of the tutors—his name is Juan—was going over a really hard multiplying fractions problem, you know where you have to cancel? Juan said, "Dude, you have good mathematical skills, but you have to show your work better". He got me to write out the greatest common factor for two numbers and how to show more clearly how to cross cancel. I guess I just never learned I had to be that organized in math.

Kathryn shared an example of how improved organizational habits helped her conquer a recurring problem with the subtraction of polynomials:

All right. Let's say you have $(3x^2 - 5x + 8) - (6x^2 + 8x + 10)$. So because you are subtracting, you have to change the signs in the second polynomial. What I used to do was just cross out the signs in the second polynomial and write the opposite, and I always messed things up. My work was just so sloppy. My professor in algebra kept after me to change the middle operation from minus to plus and in the second polynomial write $(-6x^2 - 8x - 10)$. And I started getting them right! I know that sounds really simple, but it's little things like that that helped me be a better student. (see Figure 6.2)

Tara struggled with linear equations that involved multiple steps such as operations with fractions and combining like terms. She explained how her SI leader gave her a way to remember the setup for such equations and organizer accordingly.

The steps are you do the parentheses first, then then the fractions, then you combine, then you move the variable to one side, then you add or subtract, and the last thing you do is divide. I kept getting the steps mixed up, so one day, Rob [Tara's SI leader] asked me my favorite dessert. I know this is weird, but I like cinnamon on vanilla ice cream. So, he came up with, "Please Feed me Cinnamon Vanilla All Day". Get it? The first letter of each word reminds you what step to take with linear equations. The dude's a genius. I hope he becomes a professor someday. (see Figure 6.3)

Simplify: $(3x^2 - 5x + 8) - (6x^2 + 8x + 10)$
$(3x^2 - 5x + 8) + (-6x^2 - 8x - 10)$
$-3x^2 - 13x - 18$

FIGURE 6.2
Organization of the Subtraction of a Polynomial

Solve: $\frac{1}{3}(x-2) = \frac{2}{5}(x+4) - 3x$	Apply Parenthesis (Please)
$\frac{1}{3}x - \frac{2}{3} = \frac{2}{5}x + \frac{8}{5} - 3x$	Remove Fractions (Feed me)
$5x - 10 = 6x + 24 - 45x$	Combine like Terms (Cinnamon)
$5x - 10 = -39x + 24$	Variable to One Side (Vanilla)
$44x - 10 = 24$	Add or Subtract (All)
$44x = 34$	Divide (Day)
$x = \frac{17}{22}$	

FIGURE 6.3
Acronym for Solving an Equation with Fractions and Parenthesis

Adam and Jerry also learned to develop better organizational skills; however, both students posited that their lack of organization in the past was due to apathy.
Adam explained,

> *I just didn't care; I hated math, and I just wanted to be done with the problem, so when people told me I needed to show work better, I just ignored them.*

Jerry added,

> *Maybe people told me to organize myself better in math; I don't know. I just wanted to get through a math problem the fastest possible way.*

Understanding Math Terminology. Finally!

Several students postulated that a major step to success in math stemmed from their ability to understand and apply basic math and algebra terminology. More specifically, this included their understanding of concepts such as greatest common factor, least common multiple, coefficients, terms, and polynomials. The participants further stated that the understanding of such terminology was imperative in their future success in algebra. Larry clarified:

> *I never understood the difference between a factor and a multiple. I remember frustrating teachers when they asked me what was the least common multiple between four and eight, and I would say, "four". It was actually the people in head start that finally got me to understand, and I am glad they did because understanding the difference was so important in algebra.*

Adam recalled:

> *I never understood what a term was, what a variable was, and what a coefficient*

was. Mrs. Johnson just made it so easy; I never knew it was that simple. I guess it helped that I was ready to learn.

Mike explained how his teacher, at DCC, helped the class understand math vocabulary:

> *It was on our tests, so we didn't have a choice. Professor Holton would give us a word like coefficient or variable and we had to not only define it, we had to give an example. I finally learned that stuff, and it helped me from that point forward.*

Ron specified math terminology as a math barrier in middle school and high school. Therefore, he decided to take a proactive approach.

> *I remember how those words drove me crazy, so this time I was ready. When my professor talked about factors, multiples, and coefficients, I made sure I got it I made sure I could come up with my own examples to. So, I would write $6x^2 - 5x - 3$), and I would identify the coefficients and terms and if it was an expression or an equation and why.*

Deb shared some advice regarding math terminology:

> *Look, math terms are boring; I mean they are really boring, but you need to know these words. If you don't, you will feel lost in class. Have you ever had a conversation with someone, and they keep using words that you don't know? And you're confused because you don't understand those words? Well, that's the way it is in math. If you don't understand words liker coefficients, terms, expression, equations, evaluating, and polynomials, you are gonna be lost.*

Coasters

Finally, a few of the participants admitted that they were able to complete their primary developmental math course without much effort, as they relied on previous knowledge to coast to a passing grade. Laura (FCC) explained:

> *Some of the stuff I remembered from high school, like polynomials. Sometimes all I had to do was one problem, and I remembered it, like factoring. Some stuff I didn't really get like adding and subtracting the fractions with x's [operations with rational expressions]. I didn't do too much homework, and I missed a lot of classes. I did enough just to pass by a few points; I guess.*

Laura was in the emporium model, where students can accelerate through the content.

> *Yeah, I guess; I could have finished the course early if I worked harder, but I was just lazy.*

Tom (WCC) also coasted through algebra 2 on prior knowledge. He elaborated,

> *I would look at something, like the system of equations or factoring, for like five minutes, and I remembered it. There was other stuff like the fractions with algebra [operations with rational expressions] that I remembered a little, but I didn't get right.*

Like Laura, Tom took his class in the emporium model and had the opportunity to accelerate through the course.

> *I was just lazy. I complained to anyone that would listen that I didn't belong in the class and it was too easy, but I did the least amount of work possible.*

Like Laura and Tom, Andrea found that introduction to algebra at DCC was merely a review:

> *I didn't have to study. I mean; it helped to go over problems in class, but when I saw something once, like the signed numbers or the polynomials, I got it. The class was just so boring because I didn't need all that.*

Andrea felt that she did not put forth a poor effort but not an excellent effort; it was satisfactory effort.

> *I went to class and did what I had to do; I didn't take notes; I knew enough to pass the class. It kinda caught up to me when we did word problems. I totally bombed those. Luckily, it didn't sink me in the course.*

Todd failed pre-algebra at FCC his first time. However, he posited that he was able to cobble together enough knowledge from high school math and from his first try in the course to pass during the second attempt.

> *I was actually doing pretty good for most of the class the first time till I started slacking off. So, the first part of the class was easy; I aced the first test because I just did it the last semester. I learned my lesson about skipping classes the first time, so I was better about that after the first test. But some of the stuff we did later in the course like equations came back to me all right. I didn't feel like I had to work that hard. It was more of a brush up. I didn't get the word problems at all. So, I got by with a C.*

Supplemental Instruction

I asked Dina, Ron, and Tara whether SI positively impacted their experience in algebra 1 at WCC. In general, the participants acknowledged that this was a contributor to their success.
　　Ron elaborated,

> *It was nice to have the extra time. Sometimes in class, my professor would have to move on [to another math topic] because of the time, so it was nice to have that time just to ask questions. I kept me from falling behind sometimes.*

Tara discussed the impact of her SI leader.

> *I really liked how Rob came to our classes and went over everything the same our teacher did. It gets confusing when people show you different ways to do the same problem.*

Dina shared how the SI sessions helped to relieve her math anxiety.

> *It was a relief to know that every week there was 50 minutes where I could just review and ask questions. No new material just time to catch up and practice. I also liked hearing other people's [Dina's classmates] questions, because they were questions that helped me.*

Summary

Attempting the primary developmental math class was a tumultuous task for these students as they entered these classes with a great deal of anxiety as well as the baggage of bad mathematical memories. Some students were unsuccessful in their first attempt. To be successful, the participants needed to first address personal, mental health, and time management issues as well as prerequisite skills, learning issues, and learning styles. The students also attributed their success to making meaningful connections and support networks, engaging and effective instructors, and well-developed, solid mathematical organization. With the completion of their first developmental math course, some of the participants were able to move forward to college-level math; however, others had to complete additional developmental math coursework with tougher content.

7

Math Pathways and Completing Developmental Math

Successfully completing the first developmental math course is a notable achievement, considering the large number of students who are unsuccessful. However, it is only the first step to completing a college-level math course while attempting a bachelor's degree. During their initial time at their respective institutions, some students changed their majors, whereas some continued down their original career path. Depending on their majors, students took varying directions in their math pathways to complete their goals.

The Pathways

FCC offers various pathways for non-STEM majors. Upon successfully completing pre-algebra, students can complete elementary algebra and then complete either quantitative reasoning or introduction to statistics. Students can also attempt either of the two previously mentioned courses in the corequisite model. More specifically, after passing pre-algebra, students can enroll in either QR or introduction to statistics with an additional corequisite, or booster course, that covers the basic content in elementary algebra. Students who are enrolled in STEM majors must complete elementary algebra and intermediate algebra en route to college algebra, the first college-level math course. However, after completing elementary algebra, students may enroll in college algebra with the corequisite course.

Jessica, a criminal justice major, enrolled in QR with the booster course. Larry, however, started to consider math education as a career path.

> I decided that I wanted to help other students who struggled in math and were afraid of math, like me, said Larry.

He continued as a liberal arts major, but this meant he must enroll in more math courses for his four-year degree. Therefore, Larry continued his developmental math sequence en route to college algebra, which consisted of elementary and intermediate algebra. Larry initially chose to enroll in college algebra with the corequisite course as opposed to intermediate algebra. Dan persisted as a liberal arts major, but he knew he would not be a STEM major; therefore, he enrolled in QR with the booster course. Todd persisted in hospitality, and he, like Jessica and Dan, registered in QR with the corequisite. Cindy and Laura continued as education majors, so they needed to complete the teacher preparatory course. The teacher preparatory course satisfies the college-level math requirements at FCC. This meant successfully completing both elementary and intermediate algebra.

DCC does not offer the corequisite model. Again, non-STEM majors may enroll in either Quantway or Statway. STEM majors must complete intermediate algebra en route to college algebra. Robert and Rosemary persisted as human service majors and upon completing quantitative literacy (Quantway 1), they registered for QR (Quantway 2). Harold continued as an education major, which meant the completion of the teacher preparatory courses. After completing Statway 1, Adam enrolled in the college-level portion of introduction to statistics. Andrea and Kathryn continued as early childhood education majors and needed to complete intermediate algebra, the remaining developmental math course, to register for the teacher preparatory course.

ACC is like FCC in that they offer both QR, introduction to statistics, and college algebra with the booster course. Students who complete introduction to algebra may enroll in QR or introduction to statistics with the booster course. STEM majors must complete elementary algebra and intermediate algebra to enroll in college algebra. However, after completing elementary algebra, students may register for college algebra with the booster course.

Joyce and Otis continued down their paths to a human services degree and registered for QR with the booster course. Deb developed an interest in biology, which meant more math classes; therefore, she needed to complete elementary algebra and intermediate algebra, both half-semester courses en route to college algebra. Emily decided to pursue math education, which, like Deb, meant the completion of elementary and intermediate algebra.

> *I was always so scared of math, but I had these amazing teachers who showed me I could do it. I knew, because of that, I could help others learn math,* explained Emily.

After completing algebra 1, non-STEM WCC students have the option to enroll in either QR or introduction to statistics with the booster course. STEM students must complete algebra 2 and 3 and then college algebra. Like ACC and FCC, WCC students are not required to enroll in the booster courses. They have the option to complete algebra 2 and subsequently register for QR or introduction to statistics without the booster course. Dina and Tara continued down their paths toward fine arts and human services degrees, respectively, and both registered for QR with the booster course.

Ron chose to pursue a business degree, and he registered for introduction to statistics with the booster course. In his endeavor toward an aviation degree, Tom registered for algebra 3 (intermediate algebra).

BCC does not offer the corequisite model, and they offer only introduction to statistics for the non-STEM major. As mentioned earlier, Joe pursued a degree in culinary arts and registered for introduction to statistics. Denise continued her nursing degree and enrolled in introduction to statistics. Mike decided to pursue math education and, like the others, needed several additional math courses; therefore, he enrolled in intermediate algebra en route to college algebra. Mike explained:

> *I know what math anxiety feels like. Now, that I overcame my fear, I know I can help others.*

Jerry established a curiosity for forensic science, which meant additional math classes as a STEM major; therefore, he registered for intermediate algebra so he could complete college algebra. Both FCC and ACC offer college algebra in a corequisite format.

After completing the primary developmental math courses, the participants scattered toward varying pathways. Jessica (FCC), Dan (FCC), Todd (FCC), Robert (DCC), Adam (DCC), Rosemary (DCC), Harold (DCC), Joyce (ACC), Otis (ACC), Audrey (ACC), Denise (BCC), Joe (BCC), Dina (WCC), Ron (WCC), and Tara (WCC) were finished with stand-alone developmental math courses; however, the other participants had additional developmental coursework to complete before arriving at their first or sole college-level math course. Cindy (FCC), Kathryn (DCC), and Mike (BCC) passed the remainder of their developmental courses on the first attempt. Emily (FCC) passed elementary algebra on the first try but needed two attempts to pass intermediate algebra. Andrea (DCC) and Jerry (BCC) also required two attempts, and Tom (WCC) and Laura (FCC) needed three tries to successfully complete intermediate algebra. Larry (FCC) and Deb (ACC) passed elementary algebra on their first attempt but were both unsuccessful in their attempt to complete college algebra with the corequisite course and both chose to enroll in intermediate algebra the following semester. I will convey their experiences in attempting college algebra with the corequisite in chapter 8.

Problems in Intermediate Algebra

Despite successfully completing their initial developmental math coursework, Laura, Andrea, Jerry, Tom, and Emily met adversity in higher developmental math courses. What happened?

The Coasters Crashed

Andrea, Laura, and Tom were not able to coast through intermediate algebra. Andrea recalled:

> *I fell behind really fast. I just didn't how to work hard at math—if that makes sense. We started with system of equations, and I got the two [two-by-two equations], but when we got to three equations [three-by-three equations], and I was kinda getting it, but I wasn't practicing on my own. So, when I would go back to class, I was behind because I wasn't practicing or doing homework. It wasn't like before when my teacher would explain something right away, and I got it, and I didn't have to do anything else.*

Laura, continuing in the emporium model, immediately knew that the content in intermediate algebra was unfamiliar; however, she made no attempt to change her work habits.

> *I started with functions, and I was lost right away. I don't know what to tell you; I still didn't practice outside of class. I guess I thought I would just get it somehow. The tutor worked with me a lot in class. He pretty much did my problems for me, and I just entered them in the computer, but I didn't know what I was doing,* recalled Laura.

With the expectation of a smooth ride, Tom also endeavored algebra 3 in the emporium model.

> *I just figured it would more stuff from high school that I already knew.*

However, Tom was in for a surprise.

> *Functions and all the stuff with radicals? Maybe I did them in high school, but I don't remember. I fell behind really fast.*

Like Laura, this was not a wake-up call for Tom.

> *I just figured it would start to make sense, but it didn't. Don't ask me why but I started cutting class. I guess I just didn't want to deal with it.*

Both Laura and Andrea needed to withdraw from intermediate algebra to avoid receiving an inevitable F.

> *I knew I was lost, but I still couldn't believe I had to drop the class. The thing is I never really worked hard in a math class before, but I always passed. My teacher had to convince me to drop because there was no way I was gonna pass,* said Laura.

Andrea reflected,

> *I've been lost before in math, but some way it always worked out. I always passed. There was no way out of this.*

Tom's professor suggested that he withdraws from algebra 3, but he declined.

> *I knew I was failing, but I just figured it would be all right. There were times in high school that I almost failed a class, but I somehow passed. I figured if I just studied really hard for the final, I would pass. It didn't happen.*

Harder Content, Faster Pacing, and Misaligned Modality

Emily and Jerry found the content exceedingly difficult and had immense difficulty with keeping up with the pacing of intermediate algebra. Emily shared:

> *It was just such a big step up from elementary algebra. I got the two-by-two system of equations, but I couldn't get the three [three-by-three equations]. There were just so many steps, and I kept messing up. Then, we moved on to square roots [roots and radicals], and I still wasn't getting system of equations. We were doing square roots, and I was still trying to understand system of equations, so I could pass the test.*

Emily failed her first exam by a wide margin, and she needed to take an alternative route to pass intermediate algebra.

> *I talked with my teacher and my mom. I decided to drop the class, and my mom got me a tutor, so I could work on the [intermediate] algebra on my own, so I would be prepared for the next semester.*

Emily reasoned,

> *I could get it [intermediate algebra content], or most of it, but I just needed more time and practice.*

Jerry's thoughts, regarding intermediate algebra, paralleled Emily's.

> *It was just taking me so long to understand stuff. I mean, I learned to be better organized, and I was a much better math student, but I couldn't keep up.*

Jerry, Laura, and Tom posited that the emporium model was not a proper fit for the more challenging content in intermediate algebra. Laura elaborated:

Second time through, I took the class seriously, I did my homework, I practiced, I went for tutoring, but it's so hard to learn stuff that you've never had in a computer lab and pretty much teach yourself. I never did all the crazy radicals [roots and radicals] ever in my life. I had done some quadratic equations, but not stuff like completing the square! It wasn't enough for my teacher or tutor to help me for a couple of minutes; I needed someone to explain it to me from the beginning. So, I dropped the class again, and next time I took it with an actual teacher [lecture-based instruction].

Tom shared,

Failing algebra 3 was a wakeup call for me. I really wanted to get my degree so I could be a pilot, so I decided to focus and take algebra 3 seriously, but the problem was I just couldn't learn math from a computer. I was supposed to teach myself by watching videos and doing problems on the computer and then ask my teacher if I had questions. Is that kind of backwards?

Tom decided to attempt algebra 3 in a face-to-face environment. Jerry added:

It's so scary to start something you never saw in your life like those system of equations and have to teach yourself. It's like you're already frustrated and lost by the time before you even ask for help. I wish BCC would offer this class where the teacher actually teaches students.

Although Jerry passed intermediate algebra during his second attempt, he does not feel that he learned much:

It got to where my teacher and tutor were pretty much doing the problems for me in class. I'm only saying this, because you said this will be confidential, but I don't think I deserved to pass the class. My teacher would even give me a lot of hints on tests. Maybe they felt bad for me because I already failed the class the first time.

Conquering Intermediate Algebra

The common themes for success in the intermediate algebra course paralleled some of those in the lower developmental math classes.

Recurring Themes

The participants stressed that solid organizational habits, establishing meaningful connections and math support networks as well as stellar teaching were even more imperative in this course, and some provided examples where solid organizational skill are required:

> *Adding and subtracting radicals where you have to simplify? If you don't organize and show every step neatly, you are dead,* said Larry.

Andrea asserted:

> *I learned my lesson fast. I was trying to do the word problems where you have to find the rate and distance [motion problems] and how long it took working together [work problems] by taking shortcuts. You need to label your unknown in the word problem; you need to set up tables; you need to set up your equations. If you don't, you won't get the problem right.*

Deb asserted:

> *You know when I finally got those system of equations with three of them [three-by-three system of equations]? When I started organizing my work like my professor kept suggesting.*

Harold, who only needed intermediate algebra, shared his experience,

> *When we first started doing the quadratic formula, I was like, "OK cool! I can just plug in numbers and let the calculator do the work", right? Wrong! You need to write all the numbers and signs down clearly. Then, you have to break everything down in pieces.*

Deb shared:

> *I really had trouble with adding algebraic fractions [rational expressions], and most of it was I was confusing myself. I just started writing out the steps like factoring the denominators, finding the LCD [least common denominator] and following the rest step-by-step till it was done.*

Tom discussed how his intermediate algebra teacher helped him shed poor habits.

> *I never had to organize my work in the lab [emporium model]. I just wrote stuff down wherever, so I just wasn't in the habit of doing that, but my teacher really worked with me on how to organize my work clearly so I wouldn't mess up. I spent a lot of time in his office getting help, but it was worth it.*

Some participants asserted that meaningful connections and a math support network were still imperative in intermediate algebra. As BCC does not offer a face-to-face option for intermediate algebra, Mike continued his math studies at DCC:

It was great; I was in class with some of the same students from basic algebra. I also kept up the Zoom math meetings with people from my old class and with some new people as well. We got each other through it, shared Mike. I practically lived in the tutoring center at FCC. I worked together with my friends to get through the class, recalled Cindy.

The participants attributed their success in intermediate algebra to stellar teaching, just as they did for the introductory developmental math courses.

What was great about Mr. Holton [Mike's intermediate algebra professor] was he always gave us the whole story. He didn't skip steps when explaining stuff. You need that in this class, said Mike.

I appreciate how my professor collected so much of our homework and gave us feedback, really fast, on everything. I know that's a lot of freaking work for him, but it helped me understand what I was doing wrong, explained Larry.

You wanna know how Mrs. Olive [Kathryn's intermediate algebra professor] saved my butt from failing the class? asked Kathryn. We were doing factoring trinomials, and I wasn't getting it. We were doing all the guess and check and trial and error, and I was getting so frustrated. Mrs. Olive showed me something called the A and C method. It made so much more sense, because I could just follow the steps to get the answer. The thing is you need factoring to do the fractions with x's and y's [operations with rational expressions]; you need factoring to do equations with fractions [rational equations]; you need factoring for quadratic equations. So, if Mrs. Olive didn't show me that, I wouldn't have passed. (see Figure 7.1)

Kathryn added:

Let me just say; I don't like it when teachers show you too many ways of doing the same thing; that gets confusing, but sometimes you have to show students a different way if they don't get it.

New Themes

The participants cited two factors for their success in their final developmental course that did not arise as frequently or at all when discussing the introductory developmental course: confidence and extra practice, using math software outside of class.

The participants noted that they had built up confidence in their math skills in their initial developmental math courses, and this confidence minimized previous math anxiety and allowed them to persevere.

Factor: $3x^2 - 2x - 8$

1) Multiply the A (3) and C (8) terms (disregard the signs) and obtain 24.

2) What pairs of numbers add or subtract to -2?

(6)(4)

(12)(2)

(8)(3)

(24)(1)

$-6 + 4 = -2$

3) $3x^2 - 6x + 4x - 8$

4) Now, factor by grouping:

$3x(x - 2) + 4(x - 2)$

$(x - 2)(3x + 4)$

FIGURE 7.1
The AC Method of Factoring a Trinomial.

> *By the time I started intermediate algebra, I loved math. I went from being afraid of math to realizing that I wanted to teach math for my career. I actually looked forward to math class and learning new things,* explained Larry.

> *When you're able to do stuff that you were scared of before, it makes you want to keep taking on new stuff,* said Mike.

> *It's like when you cure a fear or phobia; you just feel like you can do anything,* clarified Deb.

Some students posited that extra practice outside of class contributed to their success in intermediate algebra. Generally, all math classes require homework outside of class; however, many classes require that some or all homework is done via math software programs such as MyLab Math or ALEKS. Such software allows students to get instant feedback to a math problem and attempt a problem multiple times if they are struggling. Again, Emily had to withdraw from her first intermediate algebra course; however, she used the extra time to work with a tutor and prepare for the next attempt. Emily explained:

> *I still had access to the course shell [online software] to MyLab Math, so I would just work on those problems over and over, and it would let me do as many as I wanted. I finally got those system of equations with three equations, and I got a lot of the square roots too. I wouldn't have passed the second time without all that extra time and practice.*

Cindy shared:

> *Kids in my class were complaining about all the extra problems we had to do on MyLab Math, and it was a lot, but that's what helped me pass. I would go to the tutoring center and work on the problems on their computer over and over, and I would eventually get it. I could give you many examples, but no way could I have gotten the problems with the i's [operations with complex numbers] right if I didn't do all that practice. That's how you get math; you keep practicing. Why don't people get that?*

Laura clarified:

> *When I took the class with an actual teacher [lecture-based instruction], I saw we had to still do homework problems on MyLab Math, and I'm like, "Oh crap, not this again", but it was different, my teacher actually explained how to do the problems, and we took notes, and we did those problems after we learned how to do them. That made a big difference. It helped to be able to keep doing a problem over and over till you got it.*

Deb explained how math software outside of class made a difference in a more difficult class.

> *We used MyLab Math in my first DEV class, and it helped, but when the stuff you're doing is so much harder, you need more practice, and you need to be able to practice as much as possible.*

Intermediate Algebra Was No Cakewalk

While all ten participants passed intermediate algebra at some point, they acknowledged that the course content was much tougher than the preliminary developmental math classes.

> *How in the heck is this [intermediate algebra] a developmental class? I get this stuff is taught in high school, but this was hard,* said Kathryn.

The students provided examples of the content difficulty in intermediate algebra:

> *This was so much harder than pre-algebra and elementary algebra. I breezed through pre-algebra, and elementary algebra was harder, but there was stuff like those complex numbers and completing the square where it took me a long time to get it,* said Larry.

What makes this class so hard is the problems are just so long. I mean it can take a page to do them, and there are so many ways to mess up, and there are so many steps to remember. I don't think I ever did one of those system of equations problems without someone helping me. I couldn't do them on my own. That was the difference for me, shared Jerry.

In the first two DEV math classes, a lot of the stuff we did was somewhat familiar, so with good teachers and working together with my friends, I was able to learn, but all the stuff in intermediate algebra was new. Like, I never saw any of this in my entire life, and even though I passed, there was some stuff like the imaginary [complex] numbers that I didn't get 100%. I really couldn't get the word problems. Those rate and distance problems [motion problems] and those work problems made no sense to me. I tried and tried but could never get them set up right, recalled Cindy.

Intermediate algebra made Emily question her ability as a math student:

The math was a lot harder, but this class went so fast. It made me scared because it was still a DEV course. Could I handle a college-level class? asked Emily.

Summary

For those who required it, intermediate algebra was much more of a difficult endeavor than the introductory developmental math courses. However, the participants attributed the continued employment of solid math organization, connecting with others, stellar instruction from their professors, and using math software for extra extensive practice to completing intermediate algebra. With that, all twenty-five participants were ready to tackle college-level mathematics.

8

The End of the Rainbow

The twenty-five participants successfully completed their developmental math requirements and stood ready to tackle either their first or only college-level math course.

Jessica (FCC), Dan (FCC), Todd (FCC), Joyce (ACC), Audrey (ACC), Otis (ACC), Dina (WCC), and Tara (WCC) attempted quantitative reasoning (QR) with the corequisite. Robert and Rosemary from DCC registered for QR. Adam, from DCC, along with Denise and Joe, from BCC, registered for introduction to statistics. Ron (WCC) enrolled in introduction to statistics with the corequisite. Emily (ACC), Mike (BCC), Tom (WCC), and Jerry (BCC) attempted college algebra. Larry (FCC) and Deb (ACC) attempted college algebra with the corequisite.

Cindy (FCC), Laura (FCC), Andrea (DCC), Harold (DCC), and Kathryn (DCC) attempted the teacher preparatory course sequence. Audrey was the only student who attempted her math course in the online format. All other participants took their classes in the face-to-face modality. In chapters 6 and 7, I connected the common themes that led to both failure and success in developmental math. In this chapter, I will convey common roots of failure and strategies that led to success in college-level math.

The College-Level Courses

Quantitative Reasoning

The QR classes at FCC, DCC, ACC, and WCC cover financial mathematics (debt-to-income ratios, simple and compound interest), probability and basic statistics, proportional reasoning, linear and exponential modeling, and applications with charts and graphs. The QR course at FCC includes basic logarithms so that students can solve for an exponent. At all three colleges, the course is three credit hours and meets twice a week. At FCC and WCC, the corequisite (or booster) meets directly before the main class. At ACC, the booster course meets twice a week but on different days than the main class. The content in the booster course at FCC, ACC, and WCC parallels the content in the stand-alone Quantway 1 course offered at DCC (Appendix C). However, the booster courses pace the content so that students can learn concepts in the booster course and apply them to the QR course. For example, the booster course may cover slope and equation of the line in the classes leading up to when depreciation and resident life birth applications are covered in the QR class. This course also provides students with additional time to ask questions and review material from the main course. At all three colleges, students use a scientific calculator.

Teacher Preparatory

At both FCC and DCC, the teacher preparatory requirement is a two-course sequence.

The first course covers sequences, number theory, computation, rational numbers, and problem solving. The second course covers probability, statistics, and geometry. These courses tend to be activity-based and include some methodology for teaching elementary and middle school students as well. At both colleges, students use a scientific calculator.

Introduction to Statistics

At DCC, BCC, and WCC, the introduction to statistics course covers the methods of data description, probability, random variables and probability distribution, confidence intervals, hypothesis testing, and linear regression. At DCC and WCC, the students use a scientific calculator; at BCC, they utilize a TI-83 graphing calculator.

College Algebra

At DCC, ACC, BCC, and WCC, the content in college algebra consists of functions, polynomial functions, exponential and logarithmic functions, basic trigonometric functions, matrices and determinants, and sequence and series. Students at all colleges use a scientific calculator. At FCC and ACC, the content in the college algebra booster course consists of intermediate algebra content such as functions, quadratic equations, and roots and radicals. More specifically, students will study this content as it is needed for college algebra material. The booster course will cover basic functions as students are covering polynomial functions in college algebra. Students will study quadratic equations as they study logarithmic equations in college algebra.

The Struggles in College-Level Math

As per the requirements of the study, all twenty-five participants successfully passed a college-level math course with a grade of B or higher.

However, fourteen of the participants were unsuccessful in at least one attempt, and these students included Todd, Cindy, Dan, Tara, Robert, Mike, Rosemary, Ron, Adam, Emily, Jerry, Joyce, Laura, and Otis. Let us find out why.

Personal Issues Resurfaced

In chapter 6, a few participants shared their struggles with personal issues and how such issues hindered their success in their initial developmental math course.

Unfortunately, for Rosemary, Robert, and Adam, some issues resurfaced, and they needed to withdraw from their respective courses. Rosemary shared:

> I was excited about starting the quantitative reasoning course; I really was, but I got back with an old boyfriend, and he had a drinking problem. That just seemed to screw me up. I was basically in a codependent relationship. I started going off my medication, and I just kinda lost focus of my life.

Rosemary explained how this impacted her studies in QR:

> The same stuff was happening to me again. Without my meds, and my life such a mess, I couldn't focus. The real killer for me was probability. It's like I could understand one thing, but as soon as I had to put two or more things together to solve a problem, I was lost. I was getting frustrated and agitated in class. I couldn't keep up with taking notes and trying to understand. I kept interrupting the professor. One class, I could see and hear people laughing at me, so I just got up and stormed out and dropped the class.

Robert's problems with depression resurfaced as well. Like Rosemary, it took one mistake to lead to series of other mistakes:

> Don't ask me why I did it, but I was out with my friends one night, and I had a drink [alcohol]. That just sent me into a downward spiral. I started drinking too much, and I went off my medication for depression. It was like before; I was sleeping so much. My math class was at 2:00 in the afternoon, and I couldn't make it on time. I was always late, and I couldn't keep up or even understand what was going on. I was trying to be positive; I would try and answer questions, but I was just lost. At first, it wasn't too bad, but I couldn't get probability and those formulas.

Robert, however, did learn a lesson:

> It wasn't like before when I blamed the teacher and everyone else for my problem. I knew this was my fault, and I needed to get help again.

Adam encountered personal issues, but this time the circumstances were different. Adam elaborated,

> *My mom just started having a really hard time. She just got really depressed. Her doctor thought she never really grieved properly after my dad's death, but she was a mess, just sleeping a lot. I had to get my little brother and sister up for school and basically take care of them.*

Adam discussed how this affected his studies in introduction to statistics.

> *That course is hard. It goes fast, and there was a lot of stuff I never saw before. I just didn't have time to study or do homework, so it's like I wasn't able to really understand the stuff we were learning.*

The Underestimation of Course Content

Todd, who coasted through pre-algebra, crashed in QR.

> *Again, I passed pre-algebra without doing too much work, and when I started quantitative reasoning, we were doing mean, media, and mode and bar graphs, and I was like, "This is easier than algebra".*

Todd explained how he fell back into some bad habits:

> *It wasn't like I decided I would slack off. It was just that I was taking some hard classes like biology, and I was really into my hospitality and management courses. I spent much more time on those classes, and I didn't spend enough time on math, and I missed some classes.*

Todd was in for a shock when it came time for the first exam.

> *I looked at the practice test, and I didn't know most of the stuff. There were box plots, probability, stuff about finances, and worst of all, it was all word problems. I failed the test and the class. It wasn't the first time I did something stupid in college, and I guess it won't be the last.*

Joyce discussed her initial impression of the QR course:

> *I was surprised it was a college course. The stuff we were doing at the beginning like mean, median, and those graphs was easier than algebra. Even the probability wasn't too bad.*

This led to a similar misconception and mistake that Todd made.

> *Things were still crazy at my house. I mean; my husband was more understanding about letting me study, but still when you have a kid, you're*

working, and going to school, there just isn't enough time, so I started focusing more on my other classes. My human services classes weren't that hard, but I had to do a lot of fieldwork off campus. So, I started slacking in quantitative reasoning.

Joyce shared how this impacted her studies:

I did well on the first quantitative reasoning test, so I thought the rest of the class would be that easy. My biggest mistake was I started skipping the prep [booster] class. I thought that was a waste of time. We went over slope and equation of line [in the booster class], and I never learned that in algebra, but in the regular class we were doing all these word problems and applications with slope, so I started falling behind. You need to learn the stuff in the booster class to pass the regular class, and I was skipping the booster class and not doing the work.

Joyce admitted that this was not a wake-up call.

Once I got in my head this was an easy class, I just kept thinking it would stay easy. I kept skipping the booster, focusing on other stuff, and just falling behind in the regular class.

Cindy discovered that the mathematics for elementary school teachers is anything but elementary:

I admit it; I went into this class thinking it would be easy. I thought we were just gonna learn how to teach kids adding and subtracting. I couldn't believe this was college math! And we started doing the stuff with the sets, which I remembered from grade school, so I was like, "I got this".

Like Todd and Joyce, Cindy prioritized other subjects and parts of her life over her math course:

I kept going to class, but I wasn't really paying attention; my mind wasn't in it. I didn't do homework or go over my notes outside of class, because I thought this was just an easy class. Before I knew it, I was falling behind. We started doing harder stuff, stuff with the sets but much, much harder. We were also doing the stuff with different bases, and again, because I wasn't really paying attention, I wasn't getting it.

Cindy had a similar revelation as Joyce:

It's not a good thing to go into a class thinking it's gonna be easy, especially a math class, because you're setting yourself up for failure.

Harder Content, Faster Pacing, and Misaligned Modality: Part Two

Like intermediate algebra, some of the students found the content in college algebra very difficult and the pacing of the course to be unmanageable. Emily reflected:

> *I wish; I could tell you something different, but I had the same problem in that class I had in [intermediate] algebra. I could get stuff, but it just takes me a long time. If I had lots of time, I could learn it, but the course just went too fast.*

Emily provided an example and a metaphor:

> *I never really understood what a function was. We did those a tiny bit in the last class [intermediate algebra], but it was no big deal, but in college algebra, everything was functions, and I just couldn't catch up. It was like everyone else in the class was jumping three feet at a time, but I could only take one step at a time.*

Emily needed to withdraw from college algebra, but she did not want to spend more time retaking the course. She sought advice from her academic counselor.

> *I was just so upset, because I really wanted to become a teacher and help kids with math, but my adviser said I could still be a teacher, but I could work with younger children and help them with math, so she [Emily's counselor] said I should take the math for teachers classes [teacher preparatory classes].*

Jerry struggled in his intermediate algebra course, and it did not get easier in college algebra:

> *Every day, my professor would come in and ask us a question from intermediate algebra, you know to review. Every day, I was just lost. I can't even tell you what the review questions were about because I was so confused. He [Jerry's professor] would basically yell at us if we didn't get the review questions right. He said there is no way we would pass the course. I wish he didn't yell, but he was right. I had no clue what was going on, and I dropped the class after two weeks.*

Did Jerry anticipate problems in college algebra, since intermediate algebra was such a challenge?

> *I guess; maybe I thought that because college algebra was more traditional, you know not in that lab where we teach ourselves [emporium model]; I thought I would be all right.*

Jerry decided to return to his criminal justice major, which simply requires the introduction to statistics requirement.

Like Emily and Jerry, Mike struggled with the content and pacing of college algebra. However, Mike also became overwhelmed by his professor's teaching style.

> *I had the same guy at Drummond [Community College], Mr. Holton, for the basic algebra and intermediate algebra class, and I got so used to his teaching style. He just explained everything so well and so clearly. Mr. Holton would just start from the beginning and be so thorough. He would just come down to our level; you know what I mean, but this guy, Mr. Woods [Mike's college algebra professor], it was just so hard to understand him. Mr. Holton would come down to our level. Mr. Woods expected us to come up to his level.*

Mike clarified:

> *When teaching functions, he [Mr. Woods] would use all these letters and different symbols and advanced language. Instead of making sure we were understanding it, he just lectured most of the time. I could tell the difference between a college-level and a DEV course.*

Mike explained why he was ultimately unsuccessful in his first attempt in college algebra.

> *I felt I was hanging on by a thread through most of course. It just went so fast; I never really felt like I was getting it. It didn't help that I didn't do those math sessions on Zoom with my friends, like I did in algebra. We got to that matrices section, and it was like learning a different language. That was my breaking point.*

After a smooth ride through introduction to algebra, Ron found it difficult to keep pace with the booster course and his introduction to statistics class.

> *Don't get me wrong; I worked really hard in algebra 1. I took it seriously. But it went so much slower compared to statistics. It probably helped that I had some of the algebra before, but everything in statistics was new, and it went fast. We had to learn new stuff in the booster class and then apply it right away to the college-level class.*

Ron continued to drift as the course progressed.

> *My problem was there was stuff in the booster class like those sets and equations with the radicals that I had never done before. I was also having a lot of trouble using the calculator. In statistics, you need to learn the math and how to use the calculator. I don't know which one is harder.*

Ron ultimately withdrew from course. Like Ron, Both Larry and Deb struggled with the pacing and demand of attempting college-level math (college algebra) with the booster course. Larry clarified

> I was lost from Day 1. We started doing functions in the booster class, and I kind of got it, but we had to apply stuff right away to the college algebra class. It's like I had to learn basic functions really really fast in the booster class and use them in polynomial functions right away in the regular [college algebra] class. I couldn't keep up.

Deb concurred:

> What was frustrating was my professor would explain something in the prep [booster] class, and he would expect that we just learned it, like it would just click right away.

Deb provided an example:

> We learned quadratic equations in the prep class, and they were tough. I was having trouble a lot of trouble getting them. But then it got worse in college algebra. We learned about logarithms, which we were even harder. Then I had to use quadratic equations to help me solve log [logarithmic] equations. It wasn't happening.

Larry summarized:

> Taking the booster class and the college algebra class was like taking two hard math classes at the same time. Intermediate algebra and college algebra are two hard classes. I love math now, but I was spending hours a day on both classes, and I wasn't going anywhere.

Both Larry and Deb withdrew from their college algebra and booster classes several weeks into the semester. Both participants also felt they received poor advise from their academic counselor.
    Larry clarified:

> Look, I know I make my own choices, and I'm responsible for my own choices, but my [academic] adviser told me I should skip intermediate algebra and take college algebra and the booster. She was like, "Oh you can do it, and you'll finish faster too". She didn't tell me how hard it would be.

Deb concurred:

> My counselor had me all excited about skipping intermediate algebra and going right into college-level math. He didn't tell me how hard it would be or how hard it would be to keep up with college algebra. I went back and told him,

"You need to be real with students about how hard this is, man". I feel like he gave me some sales pitch about taking two classes at once. What can I say? He gave me bad advice; I just chose to take it.

As Dan's work hours increased, he decided to accommodate his schedule by taking QR and the booster course in an online format, but he missed the comradery with classmates:

I didn't realize how much I liked working with people in my pre-algebra class. We would get together before and after class and helped each other so much.

Dan elaborated on his struggles in the online format:

My professor set up these forums where the students could talk to each other, but it just wasn't the same. Writing to each other just wasn't the same as meeting and talking with each other. When I had a question, my professor would meet with me using Zoom, which was nice, but that wasn't the same as everyone, you know the professor and the whole class, talking with each other. That's what helps me learn. I could just tell in the first week that this wasn't gonna go well. I just felt out of place and isolated. I dropped the class, and I knew I needed a regular class the next semester.

While Dan sought a modality with more interaction, other participants felt that their college-level classes courses contained too much group work. This was particularly the case when group-based instruction was utilized as the primary instructional method. Joyce admitted that she did not put forth the effort required for the QR course, but she also found the pedagogical approach to the course very baffling:

My professor never really explained anything. We were put into these groups and we had to figure stuff out ourselves, and then the groups would have to explain to the class.

Joyce explained how this hindered her learning:

Again, I didn't put the time into the class that I should have, but it's hard to learn from each other. At first, with the graphs and the probability, it wasn't too bad, but when we got to finances and the compound interest, it was impossible. As a group, we would just struggle with stuff. My professor would come around and answer questions, but he never really explained how to do stuff.

In contrast, Otis felt that he put forth full effort in the QR course, but the group-based instruction derailed his potential success.

It was a mess. The groups in the class just couldn't get it together. You had people that weren't pulling their weight. They were on their phones and stuff, and the rest of us didn't know what to do. So, my professor kept rearranging the groups; he would put different people in different groups and it still didn't work. It was frustrating, and I was confused, especially when we got to probability. I didn't remember a thing about probability, and I was lost. My professor kept encouraging us to work together. He kept asking us questions, but he never just explained anything.

Otis decided to go to ACC's tutorial services for assistance, but this did not work.

The tutors there really weren't familiar with that course, said Otis.

In his developmental math classes, Otis asserted that connecting with his peers and forming a support network was a major piece in his success, yet group-based instruction hindered his success in QR. Why?

I need structure when learning math. If I don't know something, I need someone to explain to me in a very thorough way. Then, it helps to work with people I'm comfortable with. It doesn't work when you're forced to work with people and have to teach each other, especially when it's just chaos. It was crazy. I dropped the class before the first test.

Laura encountered frustration in her initial teacher preparatory course regarding group-based instruction:

I can't learn math by myself or working with other people. I need someone to teach me really well.

Here is an example of where group-based instruction hindered Laura's learning:

When did we ever learn different bases? Like working with numbers in base seven and base eight? My teacher put us into groups, and we had to figure out different bases while playing some game. What the heck? And that's just one example. The whole class was like that.

Laura, however, did understand that the group-based instruction was an essential part of a teacher preparatory class:

Look, I get it. We need to learn to work together to learn to be teachers, and I liked working with people with the blocks [base ten blocks] to practice teaching adding and subtracting and stuff, because that's what we're going to do as teachers someday, but the stuff with the sets and different bases and the arithmetic sequences, I never had that, and I needed someone to teach that to me. Why is that so hard to understand?

Tara became frustrated with the extensive use of group-based instruction in her QR course but also with the content within the group assignments:

My professor had us do everything in groups, which made it hard, but it was the types of problems he would give us. He would ask us questions like "How many balloons would it take to fill up the classroom?" Or "How many times can you say the alphabet in 24 hours?" He would give us some numbers, and we would have to work together in our groups. Most of the time we were lost and confused. We spent so much time on those problems, and I don't even think we got an answer.

Tara is referring to Fermi questions. These are questions that seek a rough estimate of a quantity. However, this quantity is usually difficult or sometimes impossible to measure. Fermi problems encourage critical thinking. Tara's first attempt in QR was a step backward in her math endeavor:

The class was a mess. We spent so much time on those crazy questions, we didn't have enough time for other stuff like probability and those finance questions. I didn't even know what I supposed to be doing. I dropped the class before the first test because I was so lost, shared Tara.

Even Audrey and Jessica, who successfully completed QR on the first attempt believed the group-based instruction was utilized too much. Audrey registered for the online QR class with the same group of friends that she met in the introduction to algebra class. While arranging the groups, their QR professor even allowed them to work together. However, Audrey had mixed feelings about this class.

I loved working with my friends again, and we supported each other again, but it was different this time. Our professor expected us to learn the class topics, like probability and all those finance problems, in our groups.

Audrey clarified:

In the [introduction to] algebra class, Mr. Dixon always explained the material, and he went out of his way to make sure we understood it. I'm just glad I had my friends to work with; otherwise; I'm not sure I would have passed this class. We got each other through the class, but I think I would have learned more if my teacher explained stuff like Mr. Dixon did.

Jessica added:

It seemed like all my teacher cared about was the groups working together well. Every day, it was about the groups, the groups, the groups. I had to teach myself a lot in the class. It seemed like she cared more about us working in groups than if we learned stuff.

The Midsemester Slump

Several participants shared that they hit a slump midway through the semester. Again, some of the students relayed that there were other factors such as difficult content, pacing that was too quick, and personal issues, but those issues plus heavy course loads and work issues led to fatigue and a loss of concentration and focus. Joyce clarified:

I was excited to start the semester; I had a lot of energy like I always do, but going to school, raising family, and working just takes its toll on you. That booster class was a lot more than I thought it would be. It's just the extra time. Between the booster class and the regular class, it was math four times a week. By eight or nine weeks in, I lost my drive and energy.

Jessica and Dan were successful in their first and second attempts, respectively, of QR, but both students postulated that their concentration and effort waned midway through the semester and consequently, their overall grade was not as high as it could have been. Jessica explained:

I was stoked about the about the booster class, because I would get done with my math quicker, but with the booster and the regular class, I was in math for three straight hours a day.

Jessica started attending the booster class sporadically.

I just needed the time to do homework for other classes, but it hurt. The booster class was a chance to get a head start in the regular class and do extra practice.

Dan shared his thoughts:

The semester is like a race. You start off fast, but you run out of steam, and you just don't try as hard when you get tired, and it's hard to sit through three hours of math when you're tired.

Ron articulated that it is a challenge to maintain time management skills as the semester progress:

When the semester started, I was really good about setting up time to study and do homework, but as the semester went on and things got busier, I got really bad with time management. It was hard to keep up. I wasn't going over my notes or doing homework. Taking both classes [the booster class and introduction to statistics] males it really hard to to manage your time because you are in math class for so long and have so much work.

Cindy admitted that she did not take the teacher preparatory course as seriously as she should, but the chaos during the middle of the semester did not help:

By then [midsemester], I was falling behind, and I knew I needed to get my butt in gear, but it was too late. I had to keep up with my other classes; I had to work; I wasn't sleeping enough anyway, and I was just tired.

College algebra was a struggle for Mike, but midsemester fatigue was the breaking point.

We got to matrices, and I just couldn't do it. I was trying so hard to keep up, but with other classes, working, and family, I just didn't have the brain power to try and get it, said Mike.

Denise, who passed introduction to statistics but struggled through the midsemester slump, summarized it well:

The middle of the semester is the toughest time. That's when all your classes seem to have the most work. You're tired, but you have a ways to go. You don't have the energy you have at the beginning of the semester, but you don't have the drive at the end of the semester to just finish.

Denise also acknowledged the frustrations of faculty members during the middle of the semester.

I had a couple of my teachers yell at us that we weren't trying as hard as we used to and we needed to focus. I was thinking, "I'm trying, but I'm just so tired, and I got so much going on. I'm doing the best I can".

Success in College-Level Math

If not on the first attempt, all twenty-five participants eventually succeeded in a college-level math class. Just as they did in their developmental courses, some participants needed to sort out issues in their personal lives before attempting their college-level math courses again.

Fortunately, Robert, Rosemary, and Adam were able to straighten out their personal lives. Rosemary broke up with her boyfriend and restarted her medication. Robert resumed his medication and therapy, and a relative came to stay with Adam's family while his mother got help. The next section explores the common themes and strategies that were connected to the participants' success.

Real-Life Applications: Math Actually Made Sense!

The students who completed the QR course, introduction to statistics, and even the teacher preparatory courses attributed their success, in part, to be able to relate math to the real world. Jessica expressed her thoughts regarding QR:

> I loved this class! I just loved how so much of this class related to the real world. I loved the stuff about front-end DTI and back-end DTI. I was ready to move out of my parent's house and into my own apartment. This stuff made me understand how I could afford to live on my own.

Otis, in his second attempt, echoed Jessica's thoughts:

> I liked learning about probability once I understood it. I'm a big sports fan, and there's so much stuff from probability that relates to sports. It held my interest, which made me want to learn more.

Dan shared:

> Just so many of those topics like probability and depreciation, and different types of [statistical] studies led to a lot of fun class discussions. You don't get those kinds of discussions from linear equations.

Todd realized that QR is a challenging course, and during his second attempt, he decided to take the class seriously. He appreciated how much of the course content could help him in his career goals:

> I went home and showed my dad [who operates a hotel] how you can use graphs and box plots to show approval ratings for a hotel and figure out if most people liked staying there or not. I could never do that in math.

While Tara disliked the Fermi problems in QR, Dina shared how these activities added depth to her education.

> As an artist, I'm a think outside the box kind of person. Don't get me wrong; I like structure in math, but it was nice to work on those problems with my classmates and talk about real life and different ways to solve those problems. I never did that in a math class. I learned so much.

As a future psychology major, Adam was able to tie content from psychology to introduction to statistics.

It was interesting to learn about the different type of statistical studies and hypothesis testing. I could see how you could use that to help design a treatment program for someone who has depression, said Adam.

In her teacher preparatory course, Andrea was able to obtain an answer to a long-standing question:

I finally understood why a negative times a negative was a positive! My teacher used these integer chips and related them to real life. That really got me interested in the class.

Andrea even recalled an example:

So, let's say you have a gym membership and they take $60 out of your checking account for four months. That's -60 times 4, right? So, they took out $240. But let's say you move and can't go to that gym anymore, but they keep charging your account for three months by mistake. So, you call, and they have to correct it. They basically reverse the fact that they took $60 from you for three months so that is -60 times -3, and you get $180. Finally! Someone made that make sense to me!

The teacher preparatory course also helped Cindy understand how a once confusing math concept could be a "real-life application".

Remember, how confused I was how a number divided by zero could be undefined? My professor in the math for teachers class finally helped me get it.

Cindy shared the example:

So, let's say you have fifty cupcakes to give to ten students in class. Each student gets five cupcakes, right? Now, let's say there are five students in the class; each student gets ten cupcakes, right? If there is only one student, he gets all 50 cupcakes. Now, if you had zero cupcakes to give to fifty students, each student gets no cupcakes, because there are no cupcakes to give. But if you have fifty cupcakes and there are no students you can't give any cupcakes away. So, you can't do the problem. That's why it's undefined. I get it.

For Kathryn, her revelation in the teacher preparatory course was more basic:

I never understood why you're supposed to regroup when adding and subtracting. I just thought it was something someone made up, but when we used those blocks [base ten blocks] to show why you had to regroup, it just made total sense. I get why you can't have more than ten blocks on one place. It's like if you had four tens but thirteen ones that would be the number forty-thirteen, and there is no such number. Wow!

This also shaped Kathryn's views as a future teacher:

> *Honestly, it helped me appreciate math more, and it made me realize that this was the kind of teacher I wanted to be. I want to help my students understand why we are doing things in math.*

Emily asserted that the relevancy of the content in the teacher preparatory courses as well as the hands-on activities helped her absorb the content and keep pace with the course:

> *Don't get me wrong; those teacher prep classes were hard, but I loved how my professor related everything to teaching students, and we used a lot of stuff in class, like hands-on stuff to help us understand. In algebra, I had no idea how we use those topics in real-life; I still don't. I think that's a big reason it took me so much time to learn the stuff in intermediate algebra and college algebra. If I can't understand how to apply it, it takes my brain longer to understand it.*

Joe, who completed introduction to statistics, summarized:

> *Math is still not my favorite subject, far from it, but it makes class so much better when your teacher can explain how this stuff relates to life. That just doesn't happen in algebra. It makes it easier to learn and study.*

The Positives of Corequisites

Earlier the participants shared some challenges regarding the corequisite model, such as keeping pace with the material and the extra time commitment; however, they felt that the booster courses ultimately contributed to their success for two reasons. First, the students were appreciative of the option to shorten their math pathway.

> *This [the booster course] saved me one semester of math. Hey, I've had a hard time just finishing my math requirements, so it was nice to take the booster class with quantitative reasoning and be done*, said Todd.

> *People don't get it. When you don't like math, having to take one extra semester of feels like forever. So yeah, finishing math in one semester was a big help*, explained Jessica.

Second, the students felt they were able to apply some of the developmental concepts to the college-level course because the information was fresh. Otis provided an example from the QR course:

> *We learned about lines and slope in the booster class and then right away we did resident live births and depreciation in the regular class. It was fresh on my mind.*

Dan added:

> *We went over how to solve for x with fractions [rational equations] in the prep [booster class]. Then in the main class we did those debt-to-income problems where you have to use that stuff. I could do it because I just learned it. I didn't have to wait a semester to use it.*

Organization on Another Level

Throughout developmental math, the participants stressed that utilizing mathematical organization was a key point to their success. However, for college-level math, the students required a higher organizational skill set. First, there was no time to learn organizational abilities; students need to already possess them and then apply. Larry provided an example from college algebra:

> *Log [logarithmic] equations can take over a page of paper to solve, and if you miss one step, you're dead, and I think we spent maybe one class on them and moved on. There are a lot of things you have to know to even do log equations, so if you don't know how to organize your work, you're going to make mistakes, lots of them.*

Cindy found this to be the case in the second teacher preparatory class when studying geometric proofs.

> *I think I realized one reason I could never get those geometry proofs in high school. I just didn't know how to organize myself in math. I was just always all over the place. Being able to write stuff from beginning to middle to end made doing those proofs a lot easier.*

Mike was able to understand matrices in his second try in college algebra, but he attributes that achievement to his introduction to algebra and intermediate algebra professor.

> *Mr. Holton taught me how to organize my work, and when you do matrices, especially that Gaussian elimination, stuff, you need to write all the steps out in a way you can understand them, and thanks to Mr. Holton, I did that.*

Second, the participants stressed that the solid mathematical organizational skills necessary for college-level math goes beyond written skills. Students must also become fluent, very quickly, with proper syntax when using the calculator or even programs such as Excel.
Adam shared:

> *Those probability formulas get really crazy. I actually thought it would be easy to use the calculator, but you need to be able to put those numbers and fractions in the right way or you get it wrong.*

Jessica agreed:

> *You know in the booster class, my professor told us we were gonna go over how to use the calculator to do the probability problems. I thought that was kinda stupid. I know how to use the calculator, but I'm glad he used that time to help us, because I would have been really confused in the regular class.*

Larry discussed the logarithmic equations in college algebra more:

> *You do all that work and get to the end, but you still need to use the calculator to simplify the logs to a decimal. And you have, sometimes, two or three logs, and you have fractions too. If you don't put stuff in correctly, you did all that work and get it wrong.*

The tutorial center at BCC had tutors who were well versed with the TI-83 calculator, and Denise felt that was integral in her success in introduction to statistics:

> *I would get really confused in class, especially with how to put stuff into the calculator, but those tutors just sat and worked with me on that. That's what was different to me in stats; you need to know the math, but you have to be good with the calculator. It's not enough just to understand formulas.*

Dan found the QR class interesting, but he struggled when using Excel to graph linear and exponential regressions:

> *I had never used Excel before, so I'm glad I didn't skip the booster classes where we covered that. It was really scary, and it was frustrating that so many other people in class go it so quickly, but I just practiced and learned how to put all those numbers in the exact way you need to put them in and all the steps to being able to tell if it's a linear model or an exponential model.*

For Ron, it was about the better organization of his time that led to his completing introduction to statistics. Ron's academic coach helped him map out specific time during the week to devote to introduction to statistics. His coach also suggested taking one less class to ease his academic load.

> *I wasn't spending enough time on my math, but I was also studying at all the wrong times. We worked it out, so I had 12 hours each week to devote to my math. That was time set up in advance where I couldn't schedule anything else but math. My coach told me to make sure I spent time on my statistics on the day of my math classes. I was learning material from the booster class and the regular class on the same day, so I needed to make sure I went over my notes and did a lot of practice problems when it was fresh on my mind.*

> *So, I would spend two hours that day on my statistics. Then, if I had questions, I would ask my professor the next day in his office hours. It made such a difference.*

I asked Ron if he considered enrolling in algebra 2 since he initially struggled with the corequisite format.

> *I thought about it, and to be honest, I should have taken algebra 2 the first time, but my adviser really wanted me to take the booster class, because it would get me to finish quicker. But after I withdrew from statistics, I didn't want to spend even more time on math. So, I figured I would just do whatever it took to get through the booster class and the regular class.*

Less Memorization, More Conceptualization

The students articulated that they could no longer rely solely on memorization of facts and steps to successfully complete college mathematics. Certainly, they needed to remember definitions to terms and memorize various procedures; however, it was not enough. Students articulated that they needed to develop a deeper and more conceptual understanding of the content to be successful. Dan shared his thoughts from QR:

> *That class is all word problems. You just can't memorize how to do those. You have to understand what's going on, like in those finance problems. You have to understand the difference between front-end DTI and back-end DTI and how they apply to real life.*

Audrey added:

> *You have to be able to explain your answers in quantitative reasoning. You solve a problem, like for resident live births, but you have to understand and explain what those numbers mean.*

Jerry had this realization during his study of probability in introduction to statistics:

> *We were doing probability, and I kept trying to memorize stuff, but it wasn't working, so I met with my professor, and he showed me how to use tree diagrams to help with probability. The problem was I trying to memorize formulas, but I didn't really get what probability was.*

Kathryn attempted to simply memorize when learning about different bases in the teacher preparatory course.

> *I figured out pretty quickly that to convert [numbers in] base 10 to [numbers in] other bases, you have to understand why you are doing it, or it won't make sense.*

Deb comprehended that memorization was not enough when contending with lengthy problems in college algebra.

> *I kind of figured that out in intermediate algebra, but when I got to college algebra, the problems were just so long, and I had to use the calculator so much. I had to understand what I was doing when I was doing it. Memorizing works when the problems are short, but not when they are too long.*

More Stellar Teaching

Just as they did in developmental math, the participants attributed their success in college-level math to stellar instruction. More specifically, the students appreciated thorough instruction to navigate difficult content. Geometry, which compromises a major portion of the second teacher preparatory course at ACC, was the first mathematical topic to click for Emily.

> *With algebra, I always had trouble because it always felt like I was missing something. I couldn't understand a new concept because I was missing something from before, but with geometry, I finally got it, and it's because Professor Ballard just started from the beginning and explained everything step-by-step. He basically started with the basics of an angle, which I never got in grade school and took us by the hand all the way. It just made sense, and that was because of Mr. Ballard. I couldn't believe that I could eventually do those hard geometry proofs, but I did.*

Like Emily, Jerry struggled throughout algebra. He explained how introduction to statistics was a very different experience:

> *Nothing against my algebra teachers, but I just never really got algebra. With statistics, my teacher started the first day with mean, median, and mode. He just went step-by-step. He was so thorough every class. The class was really hard, but it wasn't like algebra where I always felt behind.*

It was that teacher preparatory course where Harold ultimately understood where explanation points figured into mathematics:

> *My professor, Mrs. Green, started from the beginning and explained what a factorial was and how and why we use them, and a light just went on for me. This is the kind of teacher I will be someday. I am going to make sure I give my students full explanations that start from the beginning.*

In middle school, Denise found the topic of probability very confusing, and she was concerned when the topic arose in introduction to statistics, but this time it was different:

My teacher finally helped me understand with and without replacement in probability. He made it seem so easy. It also helped that he used visuals, like chocolates, to help explain with and without replacement. I also didn't have a good understanding of what probability was, but he filled in so many gaps for me.

Denise elaborated as to how her introduction to statistics teacher employed Team Jeopardy to help her master challenging probability concepts such as differentiating between mutually exclusive and non-mutually exclusive events, independent and dependent events, and the general probability formulas.

Mr. Jones [Denise's professor] put us into these teams, and he would give us an answer like those probability formulas, and our team would have to come up with a word problem where we would use the right formula. Or Mr. Jones would tell us to come up with a word problem where the answer was with replacement [use of independent events] or without replacement [use of dependent events]. This really helped me because I really had trouble understanding the difference between with replacment and without replacment and understanding what mutually exclusive meant..

Mike understood that there was a difference between developmental and college-level math; however, he still believes that higher level math can be taught comprehensively:

College algebra was still hard, but Mrs. Goldfarb explained it so much better than Mr. Woods. I just need someone to explain how to do a problem in a step-by-step manner in layman's terms. Mr. Woods would use all these weird symbols, and he would show these proofs of why something was the way it was, like in trig [trigonometry]. Mrs. Goldfarb still explained why we were doing stuff, but she did it in a way so we could understand, and then she walked us through each problem.

The students also emphasized that exemplary instruction did not simply consist of in-depth explanations, however. Exemplary instruction is also about teaching to the needs of students and ensuring that students are mastering the content on a regular basis. In their first attempt in their respective classes, Otis, Joyce, Laura, and Tara felt that the primary instructional modality, which was group-based, was misaligned with their learning style and, therefore, thwarted their success in the course. However, in their subsequent attempt, they appreciated how their instructors taught to the needs of the students.

We still worked in groups from time to time, recalled Otis. *But my teacher, the second time, actually taught the class how to do the problems,* shared Otis.

Otis provided an example:

I didn't get the box plots at all the first time I took the course, but my teacher, Professor Apa, just guided us through how to do a box plot. In fact, she did that the entire class.

Otis discussed how Professor Apa tapped into his students' interests and career goals to drive much of the context of the course:

She would have us fill out these surveys, and we would list things we were interested in, or we would answer questions about our major. She would use examples of that in class. Me and some other students said we were interested in football, so she used a lot of football examples. It was great.

Joyce shared:

That class [QR] is hard; there's a lot of stuff like finances, and those regressions that most students haven't really seen. We just need teachers to give us good instruction. I don't mind working together with classmates, but you have to teach to your students' needs. I'm glad I had teacher who did that the second time.

Tara spoke with her academic adviser regarding her initial frustrating experience in QR. Apparently, other QR students experienced the same hindrance from extensive group-based instruction. Therefore, Tara's adviser recommended a professor who would teach more to her needs. Tara elaborated,

It was totally different. We still did some group work, which was fine, but my new teacher, Mr. Klotz, would actually explain stuff. If I had a question, he would directly answer my question.

One day during an office hour, Tara shared her prior negative experience in QR with Mr. Klotz.

He [Mr. Klotz] said that the administration wants the quantitative reasoning professors to use a lot of group-work, so some of them feel under pressure to do it. Mr. Klotz said he doesn't care about the administration. He is going to teach to the needs of his students. I wish all teachers were like him.

The students appreciated how their teachers would employ various pre- and post-assessments to help their students' learning. Laura provided an example from her second attempt in the teacher preparatory class:

Before every topic, Professor Rath would give us questions, like those different bases [e.g., converting base ten to base five] to do in a group on that topic, but he told us not to stress if we didn't know it. He would use that to figure out what we didn't know, and then he would explain it to us really well, but if we understood something, he didn't need to spend a lot of time teaching it.

Several professors utilized various forms of a popular assessment technique entitled the One-Minute Paper. This is where students can explain, in roughly a few sentences, what they have learned in class and where they still need assistance (Angelo & Cross, 1993).

Larry clarified:

> At the end of every class, my teacher would pass out index cards, and he would have us write down what we learned but more importantly what we were still struggling with. In the next class, he would compile a list, on the board of the topics students were struggling with, and he would be sure we addressed each one. It was such a big help.

Emily appreciated Professor Ballard, her teacher prep instructor, utilizing this as well:

> As I said, I'm shy, so it was a great way for people like me to ask questions and get help. I loved it, because he would make sure we practiced the problems people were struggling with in class. It's a big reason I passed.

As in developmental math, the students valued faculty who found ways to engage students. Rosemary provided an example:

> After doing a problem, my professor would have us text our answers in class, and it showed up on the big screen. People who got the right answers, really fast, would get a prize. That was smart. Most of my other professors yell at students the whole time to put their phones away. This guy found a way to use them [phones] in class.

In her book, *Student Engagement Techniques: A Handbook for College Faculty*, Elizabeth Barkley (2010) discusses various ways to engage college students across the curriculum, and she elaborates on a technique entitled, academic controversy. This is where an instructor allows students to debate certain aspects of an academic topic thereby developing a better understanding for that topic. Laura discussed how Professor Fields, her teacher preparatory instructor, employed a variation of this technique when discussing alternate methods of addition, subtraction, and multiplication such as addition from the left, trading off, the scratch method, breaking up and bridging and lattice multiplication:

> Professor Fields would put us into groups, and he would assign a topic like breaking up and bridging or adding to the left, you know other ways to teach kids how to add and subtract. We would debate with each other which was the best way to teach kids and why. We would also figure out and explain how something like lattice multiplication could help certain kids learn. We got into some pretty serious arguments. I never knew that could happen from talking about math. It helped me learn those topics, but I feel like it prepared me more to teach kids.

Jerry shared an experience that intimidated him at first but later helped him:

> *I was scared to death about this at first, but my professor required all of us to do a presentation on our favorite topic in the statistics class. It was great. I did mine on hypothesis testing, and it helped me understand it better, but you know something; it was the other presentations that really helped me. It was interesting to listen to my classmates. They helped me understand some things that I had been struggling with. This one girl's lesson on confidence intervals really helped me. Sometimes you can understand something better when another student explains it.*

Like Jerry, Tom's college algebra professor utilized a practice that was overwhelming.

> *Every class my professor would call on a few people to write their homework answers on the board, but we never knew who he was going to call on till the beginning of class. He would give us a grade on how we did. I was freaked out at first because I never really did a good job on my math homework.*

However, Tom contributed this practice to his success.

> *My other math professors never really cared if I did my homework or not, so this was the first time I tried really hard on my homework. It was also helpful to see how other students did problems too. College algebra was really hard and having to present our homework was a big reason I passed.*

Joe's introduction to statistics professor kept his mind on math.

> *I know this is a common game, but my teacher would bring a small beach ball to class. She would throw the ball to someone and ask a question. The student would ask a question and throw the ball to someone else so they could answer. It kept me on my toes, man. It kept me from spacing out.*

In developmental math, some participants mentioned how their professors tried to connect with them regarding their progress in class. The students valued this as well in the college math. Mike shared an experience from his second attempt in college algebra:

> *Mrs. Goldfarb gave us each us of progress reports in the middle of the semester. She wrote our current grade down and some comments on how we were doing. Just the fact that she cared made me want to do better.*

Larry concurred:

> *After the first exam, my teacher had us each meet with him in his office to go over the test. He gave me some really good advice on how I could better, and it*

worked. But I was just amazed that he cared so much. I mean; the guy must have a few hundred students, right?

Maturity and Perseverance

Several students postulated that higher education taught them discipline and to focus and persist toward desired goals, and this skill set was imperative to navigating the rigors of college-level math. In some cases, they needed to learn from failure. Emily shared:

College has been tough, but you know something. I'm a lot tougher than I used to be. When I started school, I was scared of everything. Now, I have goals; I know what I want, and I am moving toward them. I know I've failed a couple of times, but because I'm stronger now; I keep going.

Like Emily, Tara feels that the college experience has made her stronger. She clarified,

It was upsetting when I had to drop the quantitative reasoning class, but if that happened a few years ago, I would have probably run home crying and dropped out of school. This time, I was confident enough to just realize that I needed to try another class. I knew I was a good student, and I would be OK.

Robert concurred:

I know that I've failed a lot in college, but I've learned to take school seriously and not blame other people. That's a big step in succeeding in school.

Joe shared:

My parents are really proud of me. They can't believe I'm the same person. I'd say a big difference is I believe in myself now. That's how you succeed in school and in life.

Otis discussed his new level of confidence.

I've been married twenty-five years, and my wife says she sees a difference in me. She said I'm more confident and that I set my mind to things more. I have college to thank for that. I went back to school a scared old man, and I passed the class that scared me the most.

Adam summarized:

If you're not ready or you don't take it seriously, college will kick your butt. You don't want an education? There's plenty of other kids who do.

Some participants credited their success in college-level math to simply knowing how to approach a math class. Deb clarified:

> *By the time I got to college algebra, I just knew how to be a good student. I went to class: I took notes; I asked questions if I needed to; I practiced the stuff I was struggling with, and I practiced questions for exams.*

Kathryn asserted:

> *Remember when I said I didn't know how to study when I got to college? I learned. Yes, I failed along the way, but I learned.*

Successful Preparation for Exams

Throughout chapters 6–8, I presented common themes and strategies to success in developmental and college-level math courses that emerged from the student interviews. Again, such strategies helped students to assimilate math content, keep pace with the course, and ultimately pass the class. However, since successful completion of these classes largely depends on achievement of exams, and these exams are often a source of anxiety to students, I asked the students about specific test preparation strategies that led to success on exams in developmental and college-level math classes.

Study Guides

All participants stated that their developmental math and some college-level instructors provided them with some sort of a study guide, or practice test, prior to each unit exam or final exam. These study guides consisted of examples of problems and applications that were covered in the unit, like the chapter summary of a textbook. The students asserted that it was imperative to complete these study guides and leave no question unanswered or incorrect.

> *The first time I took this class [quantitative literacy], I rushed to finish it [the study guide] the day we reviewed [for the exam], shared Rosemary. I got a lot of questions wrong, and I left some out. My teacher just went over them in class, so I thought I got it. I just thought doing them in class with my teacher prepared me. Wrong!*

Rosemary modified her approach to the study guide in her third attempt of quantitative literacy.

As soon as my teacher gave me the study guide, I did every problem. I got help with the questions I didn't get, but this time I tried them again on my own.

Laura admitted:

I'll be totally honest. I didn't take the study guides very seriously. They didn't count for a grade or anything, so I figured it was just extra work. But yeah, learned my lesson. I made sure I did every problem before the test.

Some participants cautioned that study guides are not exact replicas of the forthcoming exams but are still great review instruments. Larry clarified:

It's not the test, but it's a great way to practice for the test. I always worry that I won't have enough time for the test, like I'm gonna run out of time taking the test. I take the practice test to time myself, like how long does it take me to do the practice test? This way I can see if I need to work faster on certain problems.

Emily shared:

I always get so much anxiety before a test, because it feels like there is so much to cover for a test, so it was nice to have so many practice questions in one place. It just kinda calmed me down. That [the study guide] wasn't the only way I studied for my math test, but it was a good start.

Some of the college-level instructors did not provide students with a study guide. In his book, *How to Succeed in College Mathematics*, Dahlke (2011) stressed that while some professors utilize study guides, this is not a universal requirement for faculty. The absence of a study guide was a difficult transition, but one student found a way to make it advantageous.

I freaked out when I found out we wouldn't get a practice test in quantitative reasoning, shared Jessica. *But my professor had a great idea. He told us to make up on our own practice tests. He had us share our practice tests with our groups.*

Jessica found this method to be more effective than simply being handed a study guide.

It forced me to go back and review all the material, and it was great to go over the questions with my classmates. I really felt prepared for the test.

Reviewing and Fixing Past Errors

Several participants found that success on exams was linked to tying up loose ends. More specifically, they needed to review previous classwork and homework assignments (both online and paper and pencil) and focus on prior mistakes. Mike shared:

> *I never knew how to study for a math test; I guess that's why I was never good at math. Anyway, I went to Professor Holton [Mike's introduction to algebra teacher] and asked him for some study tips. He told me about twice a week to go through my homework questions and problems we did in class and make a list of problems I got wrong.*

This turned out to be a valuable study tip for Mike.

> *I would basically make up my own practice test, but it had the questions I needed to work on.*

Kathryn admitted:

> *Teachers told me I needed to go back over questions I got wrong, but it always seemed like extra work I didn't need to do. After failing [introduction to] algebra, I figured I better start listening to people. Now, I tell other students to do it.*

Joe pondered:

> *I don't know how I could be so stupid. If I was getting something wrong on my homework or somewhere, how did I think I was gonna get it right on the test?*

For Joe, this strategy was key to his passing introduction to statistics.

> *My stats professor didn't give us practices tests, like in [basic] algebra, so I kinda had to make up my own practice test. So, I just kept track of all my mistakes. I would go see my professor in his office hours with my questions. My mom couldn't believe how organized I was!*

Verbalizing Math

As part of connecting with and working with others, several participants discussed the importance of talking about, or verbalizing, mathematics. This practice served students well when preparing for exams. Cindy elaborated,

> *Oh, it makes such a difference to actually talk out a math problem. I could give you many examples, but that's how I finally figured out completing the square [for quadratic equations] and got it right on the test. I just kept talking out the steps with my study partner.*

Dina shared:

> *I needed to work with people and talk things out in the quantitative reasoning course. Some of the content is pretty abstract and it just made more sense to talk it out till we all understood it.*

> *When I just do every problem to myself, quietly, I'm more likely to make mistakes, but when you have to talk the math problem out, you can catch your mistakes,* added Dan.

Denise discussed how study partners are not always needed for verbalizing math:

> *I got a whiteboard for my house, and when I reviewed for a test, I stood up there and explained every problem like I was teaching a class.*

Data Dump/Drop Off

Students often complain that they have a good understanding of the mathematical content prior to an exam; however, when they receive the exam, their minds go blank as they are overcome with anxiety. However, Harold's intermediate algebra professor shared with him a test-taking technique that helped him and his classmates' test anxiety subside and focus on the exam. Harold explained:

> *Here's what you do. It's called the data dump. When you get your test, don't look it. Instead, take out a sheet of scrap paper and write down all the formulas and other stuff you need to remember. See it's not cheating, because you're doing it after you get your test.*

Harold explained how this strategy relieves test anxiety:

> *When I would get my math test, I would read the first question, and I would just forget everything. I couldn't remember my own name! But this way, I took a breath and just wrote stuff down, and then when I needed it, it was there.*

I asked Harold to provide an example how the data dump assisted him on an exam.

> *Let me give you two. First, no way I would have gotten the quadratic formula without the data dump. In the past, my mind would have gone blank, but I got my test, and I wrote the formula down, and it was there when I needed it, and I needed it. All right, so my second example was in my teacher prep class. I was having trouble keeping the formulas for arithmetic sequence, the Fibonacci sequence, and the geometric sequence straight. It really helped to just write down those [formulas] out before the test.*

Harold explained how the data dump helped prepare him for exams as well.

It also helps to practice the data dump before a test. So, a couple of days before the test, I would just be practicing the data dump. I would practice what I was going to write down when I got the test.

Tom's algebra 3 professor shared the same strategy but employed a slightly different name, the data drop off. Tom shared how this strategy continued to assist him in college algebra.

There is no way I would have gotten through logarithms without the data drop off. I was really struggling with the properties of logarithms. I just kept writing them over and over again to practice for the test and then I wrote them down when I got the test and they were there when I needed them.

Harold warned that the data dump is not a sole study strategy.

You still need to study your butt off. You need to do lots of practice problems and make sure there aren't problems you don't know how to do. This [data dump] is way to keep your brain from going blank when you get the test.

It is noteworthy that the strategy Harold and Tom described is a widely utilized test-taking practice. It is also referred to as a "memory dump".

Math on a Regular Basis

Several students stressed that a way to prepare for an exam is to stay sharp by working on mathematics every day or at least on a regular basis. Todd admitted:

Something I did wrong my first time, and even my second time, in pre-algebra, was I would wait till right before the exams to try and get my homework in and practice questions for the test. Basically, I would get tired and frustrated, and I really wasn't learning anything.

Audrey's aunt, a math teacher, kept her in line:

My aunt would ask, "Did you work on your math today?" And I would say, "No, we didn't have anything due today". She would get on me about working on my math every day. You know something? She was right. It kept me sharp, and I wasn't as scared when it came time for the test, recalled Audrey.

Deb shared a real-life metaphor.

I run a lot, and I have even started running in 10K marathons. So, in [introduction to] algebra my teacher was asking us when we did our math homework, and most of us said we did it right before class. She knew I was a runner, so she turned to me and said, "Do you only practice for a marathon once or twice a week?" And I said, "No, I run every day". And she said, "So why not do your math every day to prepare for a test?" It's so simple. I just never thought of it like that.

For Larry and Tara, completing math on a regular basis reduced test anxiety and math anxiety in general. Larry clarified:

I've been through a lot of therapy, and one thing I learned about myself is I tend to run away from things too much, and that only makes my anxiety worse. So, the best way to help my anxiety is just to confront things, you know, head-on. So in my first math class, when I would get nervous or overwhelmed, I didn't want to do my math, but when I just made myself do my math every day, it made me less scared, and I wasn't as scared when the test came.

Tara added:

Working on my math every day or almost every day just made me more comfortable with math. I don't know if this makes sense, but if I'm working on my math, I don't have time to be afraid of math.

Little Enrollment in Online Courses

When reviewing the course modalities that the participants chose, something stands out. Very few students chose to take classes online despite many opportunities in both their developmental and college-level classes. I asked the participants if there was a reason why they chose against online instructions. In general, two reasons surfaced. Several students felt that the online format would be a poor fit, as they were weaker math students who needed face-to-face instruction.

Don't you have to teach yourself in an online class? asked Emily. *No way I could ever do that.*

I couldn't learn math in a computer lab with a teacher in the room. How can I learn math completely on my own? asked Mike.

Math online? Like learning it on your own? I'm sure it's more convenient than driving to campus and stuff, but I need to be taught math, rationalized Otis.

Other participants never considered an online math course because it was not part of their past.

Math was always taught in a classroom, so I never even thought about taking it online, said Adam.

I never thought about it. I always took math in a classroom with an actual teacher, so I just automatically registered for a regular class, reflected Kathryn.

Student Advice for Succeeding in College Math

Each participant provided advice for students attempting each of the college-level math classes in the study.

QR

You need to be able to deal with word problems. This class is loaded with them, advised Otis.

If you take quantitative reasoning with the booster course, get your butt to that booster class. You're gonna need it, asserted Dan.

You need to make sure you're OK with equations with fractions and slopes and lines, said Robert.

Make sure you understand how to do percents, and you can round to the nearest percent. You need to do that really fast in this course. I had trouble with that, and teachers don't have time to explain that stuff in that class, suggested Rosemary.

Taking the booster course at the same time as the quantitative reasoning course sounds like a good idea because you will get through it quicker, but it's a lot of work, and it's a lot of math in one semester. You need to be ready for that, stated Joyce.

I'm glad I took the booster course and the main course in one semester, so I didn't have to take an extra semester of algebra, but sometimes it was too much; it was just too much time on math, and you need to understand stuff from the booster like equations and slope so you can do well in the regular class. I'm just saying; it's not as easy as it sounds taking the booster at the same time, reflected Jessica.

That class may seem simple at first, but it gets hard. It's also not as simple as algebra where you just have to solve for x or answer a question. You have a lot of word problems and really have to understand what's going on, and you have to be able to explain what your answers mean, cautioned Todd.

I don't know if this makes sense, but to pass the quantitative reasoning class, I had to learn to be a math student. I did that in [introduction to] algebra. I learned how to study and be responsible. No way, I could have gone right into quantitative reasoning and passed, advised Audrey.

This is a different kind of math. It's not like algebra where you have to memorize how to do every problem step-by-step. You have to really think and understand what your doing and apply it to real-life. Just have an open mind, suggested Dina.

I liked this class, but it's hard to take the booster class and the college-level class at the same time. It was a lot of work and hard to keep up. Sometimes I wish I had taken algebra 2 and then taken quantitative reasoning. My adviser pushed me into taking the booster class and the quantitative reasoning class at the same time. I get feeling they want to push people through math as fast as possible. Just remember, this [the corequisite model] isn't for everyone, advised Tara.

Teacher Preparatory

I thought it was kinda stupid I had to take intermediate algebra for this class, but you need to know stuff like slope of a line, equations, and quadratic equations for this class, recalled Andrea.

Take this class seriously. There is hard stuff like the different bases, and if you fall behind it's hard to catch up, warned Cindy.

This will be different than most math classes you have ever taken. You're gonna have to work together in groups to solve problems; you're gonna have to use blocks and [Cuisenaire] rods and other stuff, advised Kathryn.

Make sure you know the basics of how to add, subtract, multiply, and divide with fractions and decimals. If you don't know that stuff 100%, you are going to be lost when it comes to the activities. I know that sounds stupid, but I was kind of rusty with that because it had been so long, and I was used to the calculator. So, it helps if you know that stuff going into the class, suggested Harold.

Those geometry proofs are hard, but you can do them. If I can learn them, anyone can! You just have to make sure you understand everything about the basic geometry. You have to understand all those theorems. You also have to understand what an angle is, what a ray is and what a plane is. If you are stuck on anything, ask for help. If you don't will only get more lost, opined Emily.

We had to do presentations to the class. We had to present a lesson, which is kind of scary. I know a lot of kids in the class were surprised they had to do a presentation in math class, so just be ready, stated Laura.

College Algebra

Listen to me. You need to know functions. You need to know what a function is. You need to be able to understand examples of functions. If you don't get functions, you will fail, cautioned Mike.

You know how in algebra; you wonder when you will ever need factoring? Trust me, if you can't factor, you won't pass college algebra. Know factoring 100%, advised Larry.

That class goes fast. If you can't keep up with intermediate algebra, you won't be able to keep up with that class, counseled Deb.

If you don't understand something in class or on your homework, get help right away. It will hurt you in the next class. Everything in this class builds on what you have already done, advised Tom.

Introduction to Statistics

You don't really use a lot of algebra in statistics, but you need to be able to do long math problems with a lot of steps, recalled Adam.

Denise echoed Adam's thoughts:

When I first started stats, I was like, "Why did I have to take all that algebra? It has nothing to do with stats". But I realized I could do all those really long problems in stats because I could do all those really long problems in algebra, like adding fractions with x's [rational expressions]. I'll say this. I'm glad I took math in that lab [emporium model] because I got all that practice. That's my advice for people taking math: practice, practice, practice.

Joe concurred:

You have to be able to think through a problem in statistics. I could never do that or even cared about doing that before, but I learned how to do that when I took basic algebra.

Jerry conveyed:

I don't know if I'm saying this right, but here goes. In algebra, it was more about doing the math right, remembering the rules, and following the steps. It's still like that in statistics, but you also have to know how to use the calculator, and you have to know how to put stuff in correctly into the calculator. You also have to know stuff like Excel. So, in statistics, it's about the math and the technology. That's why you have to know how to do the math first.

Ron advised:

Statistics is much harder than basic algebra. It goes much faster. Make sure you manage your time. If you need to, get someone to help you arrange your schedule, so you have enough time to do math.

Summary

There was not one easy answer to successfully completing a college-level math course. In fact, it took some participants multiple attempts, and just as they did in developmental math, students had to sort out personal issues and study habits before undertaking the rigors of a college-level math course. Others, such as Jerry and Emily, needed to enroll in a more suitable math course. The combination of stellar teaching, the employment of higher level organizational skills, developing a deeper understanding for the content, and, for some, the ability to apply math to the modern world played major roles in their success. The students also emphasized the possession of proper prerequisite skills as well as a wealth of determination and perseverance to their success.

References

Angelo, T. A., & Cross, K. P. (1993). *Classroom assessment techniques: A handbook for college teachers* (2nd ed.). Jossey-Bass.

Barkley, E. F. (2010). *Student Engagement Techniques: A Handbook for College Faculty.* Jossey-Bass: San Francisco.

Dahlke, R. (2011). *How to succeed in college mathematics: A guide for the college mathematics student* (2nd ed.). BergWay Publishing.

9

I Need More Math Classes ... Now What?

The twenty-five participants in this study achieved, what they considered to be, an insurmountable goal in successfully completing a college-level math course. Additionally, they completed a math course that will likely transfer to their baccalaureate degree. For Larry and Kathryn, their newfound mathematical knowledge has allowed them to pay it forward to other students. Larry serves as a student tutor for the developmental math courses at FCC:

> *I love this job. I work with a lot of students who are just like I was. They're scared of math; they hate math. So, I show them they can do it. I feel like I can explain it them in a way they can understand*, shared Larry.

Although Kathryn does not intend to take any more math classes, she also tutors developmental math at DCC:

> *It's a way to make money through college, but I like it. I look at it as practice for becoming a teacher one day. I get on students who don't take their classes seriously. I tell them, "You can screw around and just take this class again, like I had to". I think they listen to me more because I'm a student, and I went through these classes.*

More Math?

Deb, Mike, Tom, and Larry needed more math beyond college algebra. During her studies at ACC, Deb became interested in studying biology. Deb intends to transfer to a university, and she will need to complete two semesters of calculus (calculus 1 and calculus 2). At the time of the interview, Deb was set to enroll in a semester of pre-calculus before attempting calculus. To satisfy his math requirement for his aviation degree, Tom will need to complete calculus 1. Both Larry and Mike decided they wanted to pursue bachelor's degrees in math education. At the time of their interviews, Mike was studying at Hollis University, and Larry was attending Marcus University. The math education degree, which focuses on adolescent and young adult education from grades seven through twelve, requires three semesters of calculus as well as advanced courses such as discrete mathematics, linear algebra, differential equations, algebraic structures, and real analysis.

Life after College Algebra

For students who intend to follow a pathway that consists of higher level math classes, calculus and beyond, college algebra is the gatekeeper course. Emily and Jerry attempted college algebra but were unsuccessful because of the rigors and rapid pacing of the course. Consequently, each endeavored on an alternate math pathway and succeeded. At the time of the interviews, Deb had not started pre-calculus. However, Larry and Mike both success-fully completed pre-calculus at their respective community colleges. Both received grades of B in pre-calculus. Larry recalled:

> It [precalculus] wasn't too bad. There was a lot of stuff we did in pre-calc that we did in college algebra. What got hard was the trig [trigonometry] and a lot of the word problems. We did a lot of these long business word problems with functions. I never totally got asymptotes either. I guess I passed because of a lot of the review stuff.

Mike shared:

> I got a B in pre-calculus, but I'm not sure I deserved that. I'm pretty sure my final average was in the seventies. Maybe my professor gave me some points for effort.

Like Larry, Mike felt there was some overlap between college algebra and pre-calculus:

> We did a lot more functions. It helped me understand functions better. Some of the things I was foggy on in college algebra, I got in pre-calculus.

However, Mike struggled with some of the same concepts as Larry:

> I really had trouble with trigonometry; I failed that entire section. I still don't understand how I am supposed to draw a picture with triangles after reading a trig word problem. In algebra, the word problems were more straightforward. You know, you let x be this, x + 2 be that and get it to an equation and solve. That just didn't happen in pre-calculus.

It took Tom two attempts to pass pre-calculus at WCC. Like Mike and Larry, he struggled with trigonometry.

Trig felt like a foreign language. Those trig formulas made no sense to me. I could never understand which of those trig identities I was supposed to use and why. The first time I was lost from day 1. I did a little better with it [trigonometry] the second time. I think just seeing it again helped, but I can't say I have a strong grasp on it.

Tom elaborated on his struggles with trigonometry.

I think a big problem for me in trig was that I was missing some basic stuff from geometry. That's what my pre-calc teacher said. When we did those law of sines and cosines word problems, I was supposed to understand alternate interior angles and how lines bisect and angle, and I didn't get it. When was I supposed to learn that stuff?

Like Mike, Tom found pre-calculus more abstract and less procedural than algebra.

In algebra, most of the time you get the right answer if you just follow something step by step. That's not how pre-calculus works. The word problems are so much harder, and you have to interpret a lot of stuff from graphs. I'm not used to that. Also, you have to draw pictures from reading word problems, and I couldn't even understand where I was supposed to start.

During my last communication with Tom, he was contemplating whether to take calculus 1.

I know I only need to take calculus, but I'm not sure if I'm going to pass. I can't afford to fail another class, and I barely passed pre-calculus. Maybe I should change to major that doesn't require calculus.

Rough Waters in Calculus

At the time of the interviews, Mike was in the middle of calculus 1. Larry had successfully completed calculus 1 and was in the middle of calculus 2. Mike mentioned that he is passing, but it has been a struggle.

This class reminds me of my first shot at college algebra where it's just over my head most of the time. My teacher spends so much time teaching all those proofs and theories [theorems]. I just try to follow along, but I'm lost, admitted Mike.

So how is Mike passing?

Sometimes, like with limits and derivatives, I can figure out how to compute some of the answers, like I did in algebra, but not always. I really struggle with a lot of trigonometry concepts. I'm still not 100% on functions either. I don't really understand inverse functions.

Mike expressed concerns about his ability to navigate through a math education degree:

Look, I'm not stupid. I know I have a lot more math classes to take, and I feel like I'm just hanging on in this class. I'm worried that I just don't know what's going on. Even if I can compute some limits and derivatives, I have no idea what they are. No clue. I just don't know how I'm going to pass harder math classes.

Larry passed his calculus I class with a grade of C, and it was challenging:

It was like twenty steps up from any math class I had ever taken. At my classes at Flores, my teachers always tried to make things easy for us to understand. You know; they would put math in plain language. It was way different at Marcus. My calc professor just came in and lectured.

Like Mike, Larry struggled with the abstract concepts in calculus:

My professor spent most of the class showing us all these proofs, and I didn't understand any of them. I even went to the tutorial center, and they [the tutors] made as much sense to me as my professor.

Still, Larry was cognizant that he was lacking some prerequisite skills for calculus.

I could tell there were things I should have known going into this class. I could give you many examples, but I really had trouble with optimization problems. My professor kept expecting us to "recognize similar triangles". I have no idea how to recognize similar triangles. There were also a lot of geometry formulas that I had never done. Another thing is the trig. I really didn't get trig in precalc, and in calculus we do all these applications with trig, so I'm even more lost.

Larry also found that even his algebra background was not completely sufficient:

What was really frustrating was it seemed like I didn't know how to factor enough. In calc, you do all kinds of weird factoring.

While Larry marginally passed calculus 1, calculus 2 was a complete nightmare:

In calc 1, there were at least some things I could get right, but I was completely lost in calc 2. I got a 25% on my first test. I didn't get any of the integration, and it seemed like everything we did had trig in it.

Larry found himself with two choices: withdraw or take an F.

> *With no hope of passing, I withdrew from calculus 2: The problem is I'm on scholarship, and if I don't keep a certain GPA, they drop me. I don't know if I should keep taking math classes. If I fail another class, they'll drop me.*

I spoke with Larry after his attempt in calculus 2. In addition to tutoring developmental math at FCC, Larry took on a job tutoring high school students where he provided one-on-one assistance for students who had struggled with various high school math concepts. Furthermore, he believed that this position would provide preparation for him as he worked toward his goal of teaching high school mathematics. However, this position served as somewhat of a wake-up call regarding his preparedness to teach this discipline:

> *I go in the first day thinking that all I'll be tutoring is algebra, and I can do that, but let me tell you; those kids were slamming me left and right with stuff I had never seen before.*

Larry provided some examples:

> *They were asking me about rotations, reflections, and translations. They had all kinds of questions from geometry like proofs and stuff. There was a question on equation of a circle and stuff about arcs and degrees. Someone had a question on linear programming. When the heck is that taught? Then, of course, they had trig questions. They had a lot of questions on word problems, and even if they were word problems from algebra, I couldn't do them. It's like I could do the word problems I had in algebra, but don't hit me with ones I haven't seen before.*

Larry and Mike were both fearful about taking math in college; however, they not only succeeded in their basic requirements but also discovered a love for the discipline, and they wanted to use their careers to help others intimidated by math. Unfortunately, both Larry and Mike found themselves struggling in the calculus sequence. While they were a small sample size from this study, it is possible that other students, such as Larry and Mike, may find themselves in a similar position. More specifically, students who successfully navigate the developmental math sequence may pass college algebra and then pursue a degree that requires higher level math courses, particularly in calculus. Therefore, I decided to expand the study to a larger set of participants. At Lester Community College (LCC), I consulted with nine faculty members who have master's degrees in mathematics and teach calculus. I also interviewed a graduate student who started his career in developmental math and is pursuing a master's in math. I did this to shed light on the following question: How can a student succeed in higher level math (calculus and beyond) with a developmental math foundation?

The Faculty from LCC

In selecting the participants from LCC, I sought those who possessed at least a master's in mathematics and who had taught calculus. I chose the community college setting because these faculty are familiar with the developmental math curriculum, as these courses are housed in the same department as college-level courses and may have worked with students who required developmental math coursework. I also obtained IRB approval from LCC and assigned pseudonyms to all participants. The following nine participants were LCC faculty: Professor Davis, Professor Harnich, Professor Olson, Professor Smith, Professor Weston, Professor Lewis, Professor Tibbs, Professor Schmidt, and Professor Price. Reggie was the graduate student working as a tutorial center manager. He began his academic endeavor in developmental math, and at the time of the interview was completing his master's degree in math. While Reggie is not a faculty member, he added to this study by sharing his experience as a returning student who completed calculus with a background in developmental math. I conducted the interviews via Zoom and through written questionnaire responses.

Common Struggles beyond College Algebra

I asked the faculty participants from LCC where students who attempt calculus, relying on a developmental math foundation, struggle in calculus.

Algebra and Trigonometry

The faculty mentioned that these students are often deficient in prerequisite skills found in algebra and trigonometry.

> *There is a lack of mastery of the fundamentals,* said Professor Olson.

Professor Schmidt shared that students commonly forget needed algebraic concepts:

> *They don't remember how to divide polynomials; they don't remember equation of the line; they don't remember quadratic equations.*

Professor Davis mentioned that while concepts such as factoring and simplifying complex expressions are covered in developmental math, they are sometimes not covered in the depth required for calculus. Professor Davis specifically cited factoring expressions involving negative exponents. He elaborated:

Such expressions are frequently produced by differentiation, and factoring them is key to using the derivatives to find maxima, intervals of increase, intervals of decrease, etc.

Professor Davis described another concept that is an issue:

We also don't put enough emphasis in remedial classes on simplifying complex expressions, especially by the method of multiplying the numerator and denominator of the complex expression by the LCD [least common denominator] of the interior fractions.

Professor Tibbs postulated that functions are a major issue when starting calculus:

Students know of functions, but they really don't know what they are, and they have a lot of gaps. On the first day of class in calculus 1, I ask, 'What is a function?' and very few students can answer.

Professor Lewis added that these students are very weak when working with graphs and any graphic visuals:

My students will recite how to find slope of equation of the line, but they struggle with interpreting graphs. Anytime, I start to explain a proof that involves the coordinate plane, I can see them start to struggle.

The faculty cited limitations in trigonometry as a problem in calculus classes as well.
   Professor Price elaborated:

Students get small doses of trigonometry in college algebra and precalculus. They are exposed to basic trigonometric functions, but they really don't get the deep conceptual understanding of trigonometry that they need for calculus.

The Missing Element of Geometry?

I asked if deficient skills in geometry limited calculus students who possessed developmental math as their foundation. My inquiry was based on Larry's acknowledgment of his deficient background in geometry. Dahlke (2011) acknowledged that topics of geometry are scarce in a developmental math sequence, yet this is a topic with which high schools struggle. Dahlke further asserted that pre-calculus and calculus courses are filled with geometry concepts and relationships. While some community colleges offer full developmental courses in geometry, most skim the surface with topics such as the circumference and area of circles, as well as perimeter and area of rectangles, squares, and triangles. Other topics may include volume and a brief overview of the Pythagorean theorem.

The faculty conveyed that students do not need all the concepts from high school geometry. Professors Weston, Davis, and Smith articulated that volume formulas for rectangular solids, spheres, cones, cylinders, and prisms as well as being familiar with similar triangles, equations of circles, and the Pythagorean theorem are the basic geometry essentials for calculus. However, Professors Harnich, Tibbs, and Schmidt asserted that being unfamiliar with geometric proofs can still be disadvantageous for calculus students. Professor Harnich clarified:

> *A student who did not have a geometry course has missed out on a lot of practice writing proofs. This means that the student is less familiar with making a formal argument where conclusions are based on established truths.*

Professor Tibbs added:

> *Calculus is not about rote procedures. To get a deep understanding of calculus concepts, students really need to understand proofs and more importantly, establishing a formal argument. Those who have never had to do that in geometry lack that understanding.*

Professor Schmidt provided more lucidity:

> *Think of it this way; algebra, especially the concepts in developmental math, tends to consist of deductive reasoning. It's very top-down. You start with a question and follow a standard procedure to the answer. As math progresses, students need to use more inductive reasoning where it's very bottom-up. You have to use mathematical concepts to make an argument. So yes, students who have done that prior to calculus have an advantage.*

Professor Smith shared that students often do not make the connections with geometry.

> *They see numbers and geometry as two different concepts. You miss out on geometry concepts, such as the structure of a line or an angle when you don't see how they relate.*

The Ability to Make Connections and Difficulty with the Abstract

A major issue that former developmental math students face is their inability to connect concepts. Professor Davis asserted:

> *They [the students] focus on the calculus procedures more than the concepts.*

Professor Weston added that students tend to see such concepts as a "pile of objects, not related to each other". According to Professor Smith,

There is too much reliance on memorizing rote procedures.

Professor Lewis explained:

> *When I teach limits and derivates, students are mostly concerned with how to compute them. They make no effort to try and understand them or how they relate to each other.*

Another major insufficiency, especially for former developmental math students, is their inability to navigate through more abstract concepts. Reggie elaborated:

> *Calculus was just completely different from developmental math and college algebra. I could always just see how we got answers so easily before, but when it came to finding the derivative or integrating something, I just couldn't see it.*

Other examples of such concepts include the understanding and interpretation of proofs as well as more complex graphs of functions. Professors Lewis, Tibbs, and Price are former high school math teachers and compared the foundations of a high school math background to a developmental math background.

> *High school math, and I'm talking about the track for potential college students, is rigorous and deep. You cover the procedures and the formulas, but you also go really deep into topics,* explained Professor Lewis.

I asked Professor Lewis to provide an example of this.

> *I could think of many examples, but here are a some: In our DEV courses, we get into the basics of linear inequalities, but in high school math, we would get into much harder examples; we would cover the vertices of a solution set; we would get into quadratic inequalities, and we would get into linear programming. We would also go deep in concepts like hyperbolas and ellipses. There is a lot more work with charts and graphs and interpreting graphs. We also cover a lot more word problem applications.*

Professor Schmidt added:

> *It wasn't just that we went deeper into concepts. Students had math every day for four years, so there is time to really practice concepts.*

Professor Tibbs summarized:

> *The biggest difference between high school math and developmental math is in high school students really get to see how algebra, geometry, trig, and even statistical concepts are integrated. In developmental math so many concepts are*

taught in isolation. Oh, and they get in-depth instruction on trig in high school. They get exposure to trig every day. Lots of students who take DEV math and college algebra don't even take a full trig course. When they see trig for the first time in college, it's really hard for them, and it's like a crash course.

Professor Schmidt provided an analysis:

High school math can give students a really deep and broad understanding of math concepts and that gives them the tools to understand more abstract concepts. That is why so many students transition seamlessly into precalculus and then to calculus. In developmental math, we teach them the necessary formulas, procedures, and basic concepts to succeed in college algebra or some introductory college math course.

Suggestions for Students

What suggestions do these faculty members have for students such as Tom, Larry, Mike, and Deb, who are attempting higher level math with a developmental math background?

Required Coursework

Several participants stated that students must master the prerequisite courses.

Master all of the precalculus courses: trigonometry, college algebra, and developmental math, clarified Professor Olson.

Besides college algebra and precalculus, try and take a full trigonometry course. This will give you the deep understanding for trig that you will need, said Professor Tibbs.

Reggie concurred:

I tell students, don't take college algebra with some trig or precalculus with some trig; you need to trig to be successful in calculus.

Professor Schmidt suggested:

Even if they aren't required, take the teacher preparatory courses for elementary and middle school students. These courses focus on number theory and other abstract concepts that are helpful in higher level math.

Eat, Sleep, and Breathe Math

Students who attempt higher level math courses, such as calculus, with developmental math foundations have weaker math foundations than those who completed four years of math in high school to prepare them for college. Therefore, these students need to put in a great deal of extra work. Again, the student participants attributed completing extra practice problems to their success in intermediate algebra. However, this was much more the case for these students.

Reggie expanded,

> *I just practiced a lot. I did every problem in the section whether they were assigned or not. That's how I got through it.*

Professor Harnich advised:

> *Do more homework than assigned. Keep practicing until you have mastered the concepts not just finished the assignment. This may mean that you need to work more of the less complex examples before you are ready to try the more complex examples.*

The faculty participants also stressed getting used to reading college textbooks.

> *Read mathematical texts. Ask yourself why the authors are presenting the material in the way that they do. Ask yourself how the material relates to the entire section/chapter/course,* suggested Professor Harnich.

> *In developmental math most students don't real their textbooks. Between really good teaching, doing enough practice problems, and the material being basic enough, they don't need to, but for higher level math, students need to be able to read and study from a college textbook. It's an imperative part of the learning process,* concurred Professor Tibbs.

Professor Lewis explained that reading mathematical textbooks will help students get used to the language of calculus and abstract math. Both Professor Smith and Reggie articulated that terminology such as "if, then" and "converge" and "diverge" can be barriers.

Professor Weston advised that students need to shift from the mindset of passing the class to thoroughly understanding the content. She suggested,

> *Do not be satisfied with almost understanding; think the material through until you thoroughly understand it, and also ask a lot of questions.*

Reggie concurred,

> *You have to stop focusing on just finishing the course or the degree and really focus on the math problems in the course.*

Several participants asserted that students must build a deep conceptual understanding of these topics, and that includes more abstract thinking. Professor Olson elaborated:

> *My advice to the student would include the cultivation of an appreciation for the more abstract qualities of mathematics in general, as abstraction is predominating in calculus and most higher level math courses.*

How can students accomplish this? Professor Harnich suggested:

> *Even if you are able to find a solution to a problem, don't simply move on to the next problem. Try to find a second or third way to solve the same problem. Also, get into the habit of drawing pictures, or looking at pictures, of situations to help you solve problems.*

The students in this study stressed the importance of verbalizing math in introductory college-level courses. Professors Harnich, Price, and Weston emphasized this as well for higher level math.

> *Talk mathematics with others. Make sure that you are using mathematical terms Correctly,* suggested Professor Harnich.

> *Show a classmate how to do a problem and talk it through. You will learn as well,* advised Professor Price.

> *Successful students often explain and re-explain all the new material to themselves and their classmates and relate it to all the previous knowledge. This helps them see how concepts relate to each other and not just a pile of objects in their heads,* said Professor Weston.

Summary

Like the other participants, Larry, Tom, Mike, and Deb set out to complete their college-level math requirements. However, each chose a career path that involved the completion of higher level math courses. Unfortunately, Tom struggled in pre-calculus, whereas Larry and Mike found themselves floundering in calculus.

This chapter sheds light on the struggles former developmental math students may face in higher level math. There are also suggestions for these students for preparing for higher level math, which is based on feedback from those experienced in both developmental math and calculus. Students who rely on a foundation of developmental math and college algebra may be at a disadvantage compared to students who completed four years of college preparatory high school math. Completing high school math provides students with a deeper and more abstract knowledge of math. Therefore, former developmental math students must work harder and find ways to enrich their mathematical knowledge base to prepare for higher level math. Examples of this include taking additional courses in geometry and trigonometry. Students will also need to complete extra practice and acclimate to reading their textbooks.

Reference

Dahlke, R. (2011). *How to succeed in college mathematics: A guide for the college mathematics student* (2nd ed.). BergWay Publishing.

10

Lessons Learned in the Aftermath

After examining twenty-five participants' early struggles in math, their navigation through developmental math, their eventual success in college-level math, and for a few, their endeavor beyond college math, this chapter will examine the lessons learned regarding breaking the community college math barriers. Before doing so, however, I am going to define three terms that will be used in this chapter.

The term *mathematical maturity* is employed widely and informally throughout mathematics education. In fact, there is not one formally recognized definition. Therefore, I am going to define mathematical maturity (or math maturity) as the ability to comprehend and analyze mathematical concepts; furthermore, it is the ability to understand the prerequisites, or what is needed, to comprehend and analyze such concepts. More specifically, a student with math maturity does not simply possess the ability to pass a math exam or even a math course; he or she has the aptitude to understand what prior concepts are needed to comprehend new material. Moreover, this student appreciates the time and dedication required, both in and out of class, to studying math as well as the organizational skill set to succeed in math. As an example, there were participants in this study who coasted through their developmental math courses without putting much time and effort into the course and consequently struggled in subsequent courses. Initially, they did not possess the math maturity necessary for a college-level math course; however, they developed math maturity. Math maturity can develop and evolve as well. To succeed in calculus, for example, requires a different kind of math maturity than for developmental math.

Student responsibility is another widely and informally used term. I am going to cite the concise definition offered by Jamestown Community College in Jamestown, New York, "Student responsibility occurs when students take an active role in their learning by recognizing they are accountable for their academic success. Student responsibility is demonstrated when students make choices and take actions which lead them toward their educational goals" (Jamestown Community College, 2020).

Politics is a widely used term with varying definitions. In this chapter, I am going to use Hanson's (2003) definition of politics as the competition for resources. More specifically, acting in the interest of politics is doing what is necessary to obtain various resources (e.g., money).

Lesson 1: Gaps Are Ingredients for Failure

Whether it was in middle school, high school, developmental math, or college math, a common cause for student struggles, as supported in this study, was gaps in their mathematical knowledge base. Not understanding math content can hinder student learning when such content is a prerequisite for future material. For example, proficiency in common multiples and common factors is necessary to understand fractions. Comprehension of factoring is imperative to succeed in simplifying rational expressions. Command of the properties of logarithms is essential to understanding logarithmic equations.

Gaps in the mathematical knowledge base are not limited to missing required content. As evidenced in this study, knowledge of the mathematical language is a key to student success, and the failure to understand certain mathematical language can inhibit student success. In arithmetic, this may include vocabularies such as sum, difference, product, quotient, factors, and multiples. In algebra, this includes terms such as coefficients, expressions, equations, polynomials, and functions. Students can follow an example in class; however, if they do not comprehend the terminology, they are only understanding part of the story. In fact, understanding mathematical language is imperative throughout the study of mathematics.

Gaps in the mathematical knowledge base may not be the sole reason that a student struggles with a topic, but this is certainly a common reason, and this in turn leads to frustration. When grappling with a mathematical topic, students should identify the specific prerequisite skills that are required for that topic and focus on such skills. However, attempting to grasp too many prior concepts while keeping pace with the new content often leads to students being unable to stay even with the pacing of the course and fall behind. Therefore, it is imperative to master the concepts in a previous course with prerequisite material. In this study, for example, some students were unsuccessful in college algebra because they did not fully master the concepts in intermediate algebra, which aligned with the findings of Dahlke (2011), and Boylan (2011) in that community college students attempt courses in which they are ill-prepared.

The findings of this study coincided with those from Stigler et al. (2010) and Xu and Dadgar (2017) in that some students who attempt community college math are deficient in basic arithmetic skills, which place these students at a major disadvantage for introductory algebra. Otis and Emily attempted a one-week intensive arithmetic refresher. For Otis, this was enough for him to brush up on his arithmetic skills to prepare for introduction to algebra. Emily tried the same course but was unable to keep pace with the course. Dan struggled with arithmetic as well but benefited from individualized tutoring. Jerry and Tara attempted basic algebra and struggled, in part, because of a lack of arithmetic skills. Basic calculators can perform arithmetic operations, but they cannot compensate for a lack of number sense and the comprehension of concepts such as factors, multiples, and basic arithmetic vocabulary. Consequently, Emily, Tara, and Jerry needed to devote more time, in an adult basic education class, to strengthening their arithmetic foundation. This is a reality for some students who attempt community college math. Their mathematical gaps include basic arithmetic skills, and they need more time to improve such skills before they can be successful in an introductory algebra course. However, addressing these arithmetic deficiencies is complex. Some students benefit from an intense review while others need extensive instruction.

Community colleges must continue to look for additional ways to address mathematical prerequisite skills. Corequisites have proven successful; however, this requires students to learn or refresh at the same time as they are attempting to assimilate new material. The participants who took advantage of FCC's head start program, prior to attempting pre-algebra, benefited by working to fill gaps required for pre-algebra. For students, such as Emily, Tara, and Jerry, who required extensive help in arithmetic, their respective colleges were able to connect these students with programs such as adult basic education that could meet their needs. In summation, deficiency in prerequisite skills is a common and obvious cause for gaps in the mathematics knowledge base.

Lesson 2: Student Responsibility and Mathematical Maturity Are Imperative Keys to Success

Community college mathematics has become consumed with politics. State legislatures, state advisory panels, and school administrators have been heavy-handed when imposing various initiatives to accelerate students through developmental math and to improve success rates in community college math. Some community colleges have been mandated to repeatedly redesign their developmental math sequences in attempts to accelerate students through and to improve success rates. There have been some positive outcomes, such as alternative math pathways (discussed in Lesson 5); however, the essential practice of student responsibility is missing from these initiatives.

The first piece of student responsibility is ensuring readiness for a community college math course. Again, Emily and Jerry needed extensive work in arithmetic. Additionally, Robert, Adam and Rosemary were contending with external issues that were thwarting their learning process. These students needed to address these issues with psychotherapy and medication. Furthermore, Jerry and Rosemary addressed their learning differences.

Regardless of school, these were imperative issues that the participants needed to address to improve their quality of life. Nonetheless, choosing to tackle the aforementioned issues cleared a path toward academic success.

Student responsibility and the student's level of mathematical maturity impact the choices they make prior to taking a math class. Corequisite education certainly checks the boxes for popularity. Enrolling in booster courses in unison with the college-level course allows the student to complete college-level math credit at a quicker pace. This appeases state legislatures and school administrators, and it is an attractive option for students. However, as this study has shown, attempting a booster course alongside a college-level course requires a great deal of time and work. Students must be prepared to devote more, possibly twice as much, time to their math classes with the corequisite model.

Other participants were not ready to commit to a math course, as they entered community college with a low level of math maturity, and they did not employ student responsibility. A high school diploma does not necessarily ensure college readiness. Poor attendance, which concurs with the findings of Smith (1996) and Merseth (2011), led to failure. Overall, these students did not put in the time and effort required to be successful in a math class. These behaviors were another cause for gaps in the students' math knowledge base. In some cases, students were used to passing their math classes in high school by putting forth minimal effort. Consequently, they never learned student responsibility.

Over time, the students developed their math maturity and exercised more student responsibility. How did this happen? In a few cases, their professors helped them understand what was required to be successful in math; however, in other cases, it started with the right attitude. Some participants, while apprehensive about math and unsure how to be successful students, were open and willing to put forth the time and work. Other participants had to learn from failure and recognize that the quality of life without a college education was not what they wanted. More specifically, the results of this study suggest that student responsibility and math maturity are generally not taught; they are learned.

As the twenty-first century progresses, so does student consumerism and academic entitlement, especially among Generation Z students (Harrison & Risler, 2015). This was evidenced in this study, as some participants believed that minimal to no effort in a math course would lead to a passing grade. Some students in this study refused to complete work they considered to be extra work if there was no immediate reward. One example of this was the students not participating in FCC's optional head start program. Students must understand that long-term success in math is the reward for consistent hard work. Not every task or assignment contains a prize. Unfortunately, this trend of student consumerism shows no end in sight.

In summation, students need help tackling math. They need quality instruction (Lesson 7), proper guidance, and various support systems that the college may offer. However, students must make the correct choices and put in the effort and dedication to their math courses.

Lesson 3: Math Anxiety Can Be Defeated...or at Least Diminished

Several participants entered community college with a great deal of math anxiety, and this anxiety did not develop overnight. This anxiety stemmed from humiliating experiences in math as well as intimidating teachers, which coincided with the findings of Diaz (2010). It seemed that the nontraditional students felt deeper anxiety. These findings concurred with those of Jameson and Fusco (2013). Furthermore, Boylan (2011) found that nontraditional students may be intimidated by the technology employed in a math class, which can increase the anxiety. That was certainly the case with Mike, for example, who was intimidated by the emporium model. In general, however, falling further behind in class and simply feeling lost was a great cause of anxiety. How did the participants overcome or minimize their anxiety to the point where they could function in a math class?

Community college math, especially developmental math, has experimented with many initiatives and undergone various forms of redesign. There is no reason to believe this will subside. Initiatives will be conceived, either by individuals at a community college or by external organizations, and community colleges will rush to implement such initiatives. Therefore, prior research on student's part is imperative. More specifically, when registering for a course, students should learn as much as possible about the learning modality. The emporium model, for example, was a good fit for some students in this study, as they preferred an environment where they could self-pace, have one-on-one interaction with their instructors, and focus on practice problems. However, this modality is not for everyone. Other participants found the environment and the technology daunting and confusing. It is unfortunate that BCC does not offer traditional instruction for developmental education. As I noted in chapter 2, many school administrators prefer the emporium model as it is less costly than traditional instruction, and it is a favorite of many state legislatures. The emporium model, as well as any math modality, should be utilized to best serve students; it should not exist in the interest of politics. A learning modality that is a good fit for the student can help ease anxiety; one that is a poor fit can heighten anxiety.

I will expand on this in the next lesson, but establishing connections and support networks helped lessen students' math anxiety in this study. The participants in this study were able to do this in a few ways. For example,

Harold, Mike, and Otis established study groups or partners in their respective classes. Mike worked virtually with his study partners. Larry and Cindy joined a support network at FCC's tutorial center. The participants postulated that collaborating with others made the math material less intimidating.

Lack of control was a root cause of the participants' math anxiety. More specifically, when students felt lost and confused in their math classes, they felt a loss of control and feelings of chaos, and this increased their math anxiety. However, proactive behavior toward math benefited some students and consequently eased their anxiety. This coincided with the findings of Howard and Whitaker (2011). As discussed previously, taking the initiative to research the learning modality of a class and make connections are examples of proactive behavior. Additionally, the students who took advantage of the head start program at FCC were able to start their mathematical endeavor while addressing gaps and more importantly, they were able to get a head start on the required mathematical content in a relaxed atmosphere.

For a student who has math anxiety, starting that first class is daunting; therefore, students should seek various methods to engage in proactive behavior. For example, students could research the introductory content in their course and view online videos as well as obtaining tutoring. Community colleges should consider offering programs such as FCC's head start, and students should exercise student responsibility and take full advantage of them. In summation, getting ahead and gaining control ease math anxiety.

While examining this study, the participants' anxiety decreased as their level of math maturity increased. Can a student with a high level of math maturity still experience math anxiety? Absolutely, but as the students gained an understanding of how to approach and prepare for a math class (addressing prerequisites, time management, putting in the required effort and work), they became more at ease with the discipline. Achieving success in math also lessened their anxiety going forward. This was beneficial to the students who encountered failure in subsequent math classes even after finding success. This new brand of confidence allowed them to persist and eventually complete their college-level math requirement.

Lesson 4: Connections Should Be Made, Not Forced

Again, several participants stressed that establishing meaningful connections with peers was an imperative part of their success. Starting community college can be intimidating for some students, and establishing such connections allowed the participants to feel less isolated and provided them with a sense of belonging. This was the case for Cindy who struggled with shyness as well as Mike and Otis who felt out of place because of their ages. Conversely, when students such as Otis and Emily noted feeling isolated in their initial course, this seemed counterproductive to their endeavor.

Another benefit of collaborating with peers is that it requires students to verbalize math. The participants mentioned that explaining math problems to their peers deepened their understanding of the content. Students who were struggling with content also were able to learn from their peers.

Collaborating with peers also contributed to the development of student responsibility. For example, Mike shared that if he did not complete his work, he added no value to the study group, and this group was an imperative part of his success in math.

How can students make connections with peers? In this study, students were able to form study groups and partners in class and outside of class. Dahlke (2011) suggested that math faculty can play a role in this process as well. Again, faculty can allow and encourage to work collaboratively in class. However, Dahlke suggested that faculty can assist students in arranging study groups. Faculty can create sign-up sheets for study groups that fit with their schedules. Since community college students have busy schedules, faculty can create virtual groups, similar to Mike's class, for students to meet.

In her book, *Community College Success,* Adney (2012) discusses the isolation that community college students face. She suggested that new students attempt to make new friends and connect with others and with the college. Adney suggested this can be done by arriving to class early and sitting with people in class who are outgoing. This may also include joining a club on campus as well. These strategies can combat isolation and help students make connections, which can enhance learning.

While making connections with peers positively impacted their success in math, the participants did not appreciate the overuse of group work. Students in the QR course were frustrated by their instructors using group-based learning as the primary source of instruction. The participants noted that this created disorder as some groups did not work well together; some students in various groups did not pull their weight, and instructors responded by simply rearranging the groups. Moreover, the participants were frustrated that this method was prohibiting them from understanding new and difficult material. Students in the teacher preparatory courses expressed similar frustration, as this was new content as well. The participants postulated that group work was beneficial in class; however, they valued initial instruction from their teachers.

In summary, working collaboratively in math is beneficial to students, and this method allows students to develop strong interpersonal and problem-solving skills, which are imperative for job experience. However, faculty should be careful about forcing too much group work on students, and clearly group-based instruction is not a proper fit for all students. As Cafarella (2020) noted, some QR faculty feel pressured by the Dana Center and Carnegie to employ group-based instruction. Faculty should be able to utilize teaching methods that best suit their students, and pedagogy should not be compromised by politics.

Lesson 5: Alternate Math Pathways Benefit Students

Traditionally, most students needed to complete long sequences of developmental math before attempting college algebra. Students who are not science, technology, or engineering majors, and therefore do not need calculus courses, can complete less developmental math courses and transition into QR or introduction to statistics. Students can also enroll in booster courses at the same time as their college-level courses, which can accelerate their math completion. This study's findings show that such pathways can be beneficial to students.

QR and introduction to statistics contain a lot of content, such as finances and probability, that students were able to relate to real-life situations. This was also the case in the teacher preparatory courses, as students were able to relate the content to their future career goals. The participants asserted that being able to relate math to the modern world provided them with motivation. Furthermore, this made the material more comprehendible as well. This aligned with the findings reported by Howard and Whitaker (2011). Assimilating content at a necessary rate is imperative to keep pace with the course and avoid gaps in the knowledge base. Being able to relate to the content and seeing the value in such content allowed the students in this study to do that.

As mentioned before, enrolling in a booster course alongside a college-level course requires a heavy time and work commitment. Students not only need to spend more time on their math, but they need to be able to learn the material and then apply it very quickly. In this study, students were able to do that in the QR and introductory to statistics course. While Larry and Deb struggled with the college algebra corequisite format, studies have shown that students who enroll in college algebra with a booster course do as well as those who enroll directly in college algebra (Smith, 2019; Vandal, 2016). Success with the corequisite model depends on the student's ability to assimilate new content and immediately and successfully apply it to the college-level course.

The corequisite model can be beneficial to students who put forth the time and effort. Students can complete their college requirements at a quicker, and less costly, pace. Also, the new knowledge gained in the booster course is fresh when applied to the college-level course. More specifically, in the traditional model, students learn content in a developmental course but do not apply it till at least the following semester. Therefore, in select cases, it is more beneficial to learn new content and apply it almost immediately.

Lastly, college algebra is not for all students. The content is not suitable for all students, and students who do not require calculus clearly benefit from other courses such as QR and introduction to statistics. The statistics have shown that students struggle with long routes through developmental math just to arrive at college algebra. In this study, we have seen both the academic and personal struggles that can derail students. Therefore, it is beneficial to have alternate math pathways for students willing to put forth the effort and demonstrate competency.

Lesson 6: But Stand-Alone Developmental Math Is Still Necessary

Developmental math has received a lot of negative attention. Politicians have grown frustrated, as they view developmental education as a drain on higher education spending. School administrators feel the discipline thwarts student success and negatively affects state funding. Students get irritated spending extra time in math courses. Developing alternate math pathways was necessary. However, stand-alone developmental math courses still have their place in higher education.

Students still need college algebra, and the fast-paced curriculum with challenging content requires a solid mathematical foundation. In this study, students who struggled in and marginally passed intermediate algebra were not successful in college algebra. More specifically, the gaps in their knowledge base, required for college algebra, were too plentiful and wide. The bottom line is students need proficiency in the intermediate algebra content for college algebra, and for some students this means successfully completing a stand-alone intermediate algebra course.

Stand-alone developmental math classes can still serve students who take the pathways to QR, introduction to statistics, and the teacher preparatory courses. First, there are basic skills that students need to keep pace in those courses. Second, developmental math courses help students develop the mathematical maturity that they need for college-level math courses. In this study, several students developed the organizational skills and work habits required for their respective college-level math courses.

Traditionally, students who placed below college algebra were required to take at least one stand-alone developmental math course, but this is not necessary for all students. Based on their mathematical ability, which can be measured by a placement test, some students can bypass stand-alone developmental math by enrolling in alternative college-level math classes as well as a corequisite model. However, for students whose mathematical skill set is too far from college-level and are not able to assimilate and apply new content through a corequisite model, some form of stand-alone de-velopmental math is imperative. In their interview with Dr. Hunter Boylan, former longtime president of the National Association of Developmental Education, Levine-Brown and Anthony (2017) reported that a one-size-fits-all approach is disadvantageous to students. Incoming community students have varying abilities and needs regarding math, and community colleges must be accommodating to such abilities and needs.

As I discussed in chapter 2, students have entered higher education academically unprepared since its inception and have needed some sort of remedial education. History has shown that remediation in tandem with college-level courses has not been enough for students, and therefore, stand-alone remedial courses have existed since the 1800s, and this need will likely not dissipate.

Lesson 7: Stellar Teaching Is Invaluable

Teachers played a salient role in the participants' math endeavor. When I inquired about the students' experiences in math prior to community college, many had vivid memories of their teachers. In some cases, negative and fearful experiences with teachers painted a pessimistic flavor for math and even contributed to the participants' math anxiety.

Several students entered college with negative experiences in math, but caring and patient teachers eased their anxiety and helped them begin their math endeavor. Examples of this included taking an interest in the students' learning and creating an environment where students could ask questions and make mistakes.

Whether it was developmental or introductory college-level math, students appreciated thorough instruction. More specifically, when faculty explained the content from beginning to end, and addressed problematic areas, students felt this put them on the path to success. This was not limited to detailed lectures or guided practice; good teaching consisted of employing methods to ensure that students understood the material. Effective teachers used a variety of methods in class to reach students. This included using techniques to identify student interests and career goals, which led to more personalized instruction. This also included assessment techniques, such as the One-Minute Paper, which teachers utilized to understand and address student needs. Lastly, exemplary teaching included quick and thorough feedback.

The students articulated that faculty who were engaging played a major role in their success. Such faculty members found ways to keep students involved during math class. Whether it was tossing a ball when asking a question or having students text their answers in class, students were engaged in learning math.

A few years ago, I was attending a dreadful conference. Like many, I look forward to attending academic conferences to learn more about effective pedagogy. However, in this particular conference, the attendees were trying to outdo each other using contemporary buzz words relating to pedagogy. This happens in higher education, as many individuals feel employing buzz words puts them in a better position with external organizations and the state for receiving funding. Finally, one gentleman stood up and described how his pedagogical practices may not seem exciting or innovative; in fact, they were rather ordinary. He asserted that his goal was simply to provide thorough and complete instruction and to engage his students. In academe, it is imperative that we continue to search for innovative ways to help students in mathematics. However, quality pedagogy should ultimately focus on the best way to teach and engage students.

Whether it is elementary school, high school, or college, teachers play a critical role in the students' mathematical journey. Student responsibility is imperative. However, teachers can help guide students on a clearer path by being patient, caring, and engaging. Teachers can help the learning process by providing thorough instruction while using a variety of techniques to meet and assess their students' needs.

Lesson 8: Higher Level Math (e.g., Calculus) Requires a Different Skill Set and Mindset Compared to Introductory College-Level Math

The primary goal for students who place into developmental math is to complete their basic college-level math requirement. Such students generally do not choose degree paths in the STEM fields, and many community colleges employ math pathways to redirect these students away from STEM paths or even the college algebra route. It is less probable that students who begin in developmental math eventually choose a path that involves higher level classes (e.g., calculus); however, as evidenced in this study, it is possible. Therefore, these students, and their needs, must be addressed.

The twenty-five participants entered their respective community colleges underprepared for college-level math. However, through their course work, both developmental and introductory college-level, they developed the prerequisite, organizational, and study skills needed to succeed in a college-level math course. The students' growth regarding responsibility and math maturity were key factors as well. Nevertheless, the skill set required to complete introductory college-level math does not necessarily extrapolate to higher level and more abstract math courses.

The required skills needed in courses such as calculus certainly parallel those needed to complete an introductory college-level course. However, students must cultivate much more of an abstract and deeper conceptual understanding of math. Examples of this include interpreting graphs and functions as well as successfully constructing and understanding mathematical arguments. Furthermore, students must have a thorough understanding of algebra and trigonometry in addition to some geometry essentials. It is also necessary for students to understand how these topics relate to each other. Success in calculus also requires more independent learning. The faculty participants from LCC asserted that students gain such skills through the rigors of college preparatory high school math. In contrast, students who begin in developmental math must work much harder and complete additional preparation to succeed in higher level math. As Professor Schmidt pointed out, these students must transition from deductive to more inductive thinking.

Community colleges should be prepared to assist these students. The pathway from developmental math to college algebra to pre-calculus to calculus may not provide students with enough of a skill set. Advisers and others should work with these students regarding the additional steps and preparation needed (e.g., supplementary coursework or practice) for success in calculus. Ultimately, students who attempt calculus with a background in developmental math and college algebra are at a disadvantage compared to those who completed, and especially flourished in, college preparatory high school math. Therefore, they need to understand the required skill set and mindset, so they are not blindsided by higher level math.

Lesson 9: Learning Math Does Not Always Follow a Hollywood Script

In the Hollywood film *Legally Blonde*, the lead character, Elle Woods, played by Reese Witherspoon, begins Harvard Law School unprepared and faces one obstacle after another. However, by the film's end, she is on stage, at her graduation, having been selected valedictorian. Her former adversary is now her best friend; her boyfriend is about to propose, and everyone is cheering. Learning and achieving success in math does not necessarily follow this script.

In this study, while the level of dislike varied, all the participants entered their respective community colleges abhorring math. It took a great deal of courage for some of the students to even attempt math at community college. Some were successful in their first attempt in developmental math;

some were not. For others, it took multiple attempts to complete developmental math. After succeeding in developmental math, some floundered in their college-level courses before succeeding. Three participants pursued math beyond college algebra, but at the time of the interviews, they were struggling. More specifically, the path to success in math is not necessarily an experience that follows a linear process. How did students who met adversity persist? For some, improving their quality of life was a motivator. Many students developed student responsibility and math maturity, which helped them improve and succeed. Others asserted that learning from failure helped them persist.

The unfortunate reality is that many students never succeed in their college-level math course or even emerge from their developmental math sequence. This study provided empirical insight as to some of the struggles these students face. Some students enter a developmental math course academically underprepared to where they require more of an education in arithmetic. Other students have too many gaps in their knowledge base and cannot keep up with the pacing of a math course. Various students simply do not possess the student responsibility or math maturity to be successful in math. As evidenced in this study, students may struggle with personal issues such as depression, alcoholism, and even homelessness. The road may be winding, and it may not follow a Hollywood script, but this study shows that success in college math is possible.

Final Thoughts

There are a lot of statistics, most of them negative, that are related to community college math. This study has shown that behind these statistics are students; they are human beings who not only struggle in math but also in real life. Student situations when tackling math are not as simple as statistics and not as simple as I assumed when I set out to teach community college math.

There have been and will continue to be initiatives geared toward improving community college math. Administrators and faculty must be wary, however, when designing and implementing such initiatives. In his book, *Good to Great*, Jim Collins (2001) asserts that initiatives that enjoy longitudinal success result from employing the flywheel concept. This concept consists of setting goals and moving toward these goals in a consistent and disciplined manner while allowing the initiative to solidify but also gradually improving the initiative. This contrasts with the doom loop concept. Employing the doom loop concept consists of embracing constant radical changes and proceeding without clear direction. The doom loop tends to produce unproductive fads as opposed to successful initiatives.

Clearly, the flywheel concepts should be utilized when implementing or redesigning math initiatives.

Can community college math barriers be broken? Yes, but it is not simple. Consider that in a lifetime, people make commitments such as dieting and exercise to lose weight and large commitments such as getting married and having children. To follow through on such commitments requires the correct mindset and dedication. First, students must be ready to make the commitment to mathematics and their education. It takes the correct prerequisite skill set, student responsibility, math maturity, discipline, and perseverance to be successful in math. Many students enter community college with a dislike and downright fear of math, and they need support. There is no catholicon for student struggles. They need caring and quality teachers; they need a support network within their community college. In some cases, community colleges must provide students with outside support for their own lives. While implementing or redesigning math programs at community colleges, administrators and others must understand that one size does not fit all students. Students have varying learning styles and preferences, and this must be reflected in course and modality design. Furthermore, student-centered pedagogy should always be prioritized over politics. Now, as a great friend and former mentor of mine used to advise his class, let's go forth and do math!

References

Adney, I. (2012). Community college success: How to finish with friends, scholarships, internships, and the career of your dreams. NorLightsPress.com.

Boylan, (2011). Improving success in developmental mathematics: An interview with Paul Nolting. *Journal of Developmental Education, 34*(3), 12–41.

Cafarella, B. (2020). Community college perspectives regarding Quantway. *Community College Journal of Research and Practice.* doi:10.1080/10668926.2020.1719940.

Collins, J. (2001). *Good to Great.* HarperCollins: New York.

Dahlke, R. (2011). *How to succeed in college mathematics: A guide for the college mathematics student* (2nd ed.). BergWay Publishing.

Diaz, C. R. (2010). Transitions in developmental education: An interview with Rosemary Kerr. *Journal of Developmental Education, 34*(1), 20–25.

Hanson, E. M. (2003). *Educational administration and organizational behavior* (5th ed.). Pearson Education Inc.

Harrison, L. M., & Risler, L. (2015). The role consumerism plays in student learning. *Active Learning in Higher Education, 16*(1), 67–76. https://www.learntechlib.org/p/159326/

Howard, L., & Whitaker, M. (2011). Unsuccessful and successful mathematics learning: Developmental students' perspectives. *Journal of Developmental Education, 35,* 2–16.

Jameson, M. M., & Fusco, B. R. (2013). Math anxiety, math self-concept, and math self-efficacy in adult learners compared to traditional undergraduate students. *Adult Education Quarterly, 64*(4), 306–322. https://doi.org/10.1177/0741713614541461

Jamestown Community College (2020). *Student responsibility statement.* https://www.sunyjcc.edu/student-experience/student-responsibilities/student-responsibility-statement#:~:text=Responsible%20students%20take%20ownership%20of,seminars%2C%20prepared%20and%20on%20time

Levine-Brown, P., & Anthony, S. W. (2017). The current state of developmental education: An interview with Hunter R. Boylan. *Journal of Developmental Education, 41*(1), 18–22.

Merseth, K. M. (2011). Update: Report on innovations in developmental mathematics–Moving mathematical graveyards. *Journal of Developmental Education, 34*(3), 32–39.

Smith, A. D. (2019). Relationship between required corequisite learning and success in college algebra. *Georgia Journal of College Student Affairs, 35*(1), 23–43. https://doi.org/10.20429/gcpa.2019.350103

Smith, D., O'Hear, M. O., Baden, W., Hayden, D., Gordham, D., Ahuja, D., & Jacobsen, M. (1996). Factors influencing success in developmental mathematics: An observational study. Research and Teaching in Developmental Education, *13*(1), 33–43.

Stigler, J. W., Givven, K. B., & Thompson, B. J. (2010). What community college developmental mathematics students understand about mathematics. *MathAMATYC Educator, 1*(3), 4–16.

Vandal, B. (2016). Coreq and college algebra. *Complete College America.* https://completecollege.org/article/coreq-and-college-algebra/

Xu, D., Dadgar, M. (2017). How effective are community colleges math courses for students with the lowest math skills? *Community College Review, 46*(1), 62–81. https://doi.org/10.1177/0091552117743789

Appendix A: Analyzing the Results

This was a qualitative study, which meant that my methods were inductive and not deductive as in a quantitative study. Therefore, the data were in words, not in numbers. In chapters 4 through 9, my findings were categorized. To come to these categories, I employed the practice of constant comparison. This is when the researcher continues to read and examine the participants' feedback and identify common themes. The process of constant comparison is completed when the researcher has achieved saturation. Again, this is when the researcher is no longer discovering any new insights from studying the data (Merriam, 2002b).

Quantitative studies recognize threats to internal validity such as selection bias, maturation, testing. Moreover, researchers employ methods to try to control these threats as much as possible. Since qualitative studies differ greatly in method and data, there are various ways to ensure validity, or in the vase of qualitative studies, trustworthiness.

For this study, I employed the strategy of member checking. Krathwohl (2009) stated that member checking involves requesting the "participants to read the researcher's report to determine whether it has portrayed them accurately" (p. 346). I emailed my participants my categorical findings with a brief description and asked if they believed their responses were grounded in those findings. While a few of the participants requested additional clarification, all the participants concurred with my findings.

A study should provide some measure of external validity. Moreover, can the results of the study be extrapolated? Therefore, I employed a thick, rich description to convey my findings. Merriam (2002a) suggested that offering a thick, rich description will permit readers to conclude how thoroughly findings between studies match and whether they can be transferred. Patton (2002) asserted that studies that are rich and thorough enlighten the reader. Additionally, Ridenour and Newman (2008) postulated that a thick, rich description can allow a reader to feel like part of the story.

Appendix B: Pre-Algebra and Introduction to Algebra Course Content

Pre-Algebra Flores Community College (Half Semester)

- A brief review of operations with fractions and decimals
- Operations with rational numbers
- Evaluating expressions
- Solving linear equations
- Word problem applications

Introduction to Algebra at Drummond Community College (Full Semester)

- Review of procedures with fractions, decimals, and percentages
- Operations with rational numbers
- Operations with polynomials
- Solving linear equations
- Lines and slope
- Factoring expressions
- Problem-solving applications
- Basic geometry (circumference, area, and volume)

Introduction to Algebra at Arnold Community College (Half Semester)

- Operations with rational numbers
- Evaluating expressions
- Procedures with polynomials
- Solving linear equations
- Problem-solving applications

Basic Algebra at Blair Community College (Full Semester)

- Review of operations with fractions and decimals
- Operations with rational numbers
- Evaluating expressions

- Solving linear equations
- Lines and slope
- Factoring expressions
- Operations with rational expressions and rational equations
- Basic geometry (circumference, area, and volume)
- Problem-solving applications

Algebra 1 at Walsh Community College (Half Semester)

- A brief review of fractions and decimals
- Operations with rational numbers
- Evaluating expressions
- Linear equations
- Slope and equation of the line
- Word problem applications

Appendix C: Stand-Alone Quantway 1 and Statway 1 Content

Drummond Community College (Full Semester)

Quantway 1 (Quantitative Literacy)

- Interpreting charts and graphs
- Review of fractions, decimals, and rounding numbers
- Mean, median, and mode
- Operations with ratios and proportions
- Linear equations
- Slope and equation of the line
- Basic business applications
- Problem-solving applications
- Introduction to Excel

Statway 1

- Distributions
- Rounding numbers
- Linear equations
- Sets
- Equations with square roots
- Statistical notations
- Predicting change

Appendix D: Elementary Algebra (All Half Semester) Content

Flores Community College

- Lines and slope
- Operations with polynomials
- Factoring expressions
- Operations with rational expressions and rational equations
- Basic geometry (circumference, area, and volume)

Arnold Community College

- Lines and slope
- Factoring expressions
- Operations with rational expressions and rational equations
- Absolute value equations
- Basic geometry (circumference, area, and volume)

Walsh Community College

- Operations with polynomials
- Factoring expressions
- Operations with rational expressions and rational equations
- System of linear equations

Appendix E: Intermediate Algebra Content

Flores Community College (Half Semester)

- System of linear equations
- Functions
- Quadratic equations
- Roots and radicals
- Complex numbers
- Problem-solving applications

Drummond Community College (Full Semester)

- Operations with rational expressions and rational equations
- System of linear equations
- Functions
- Quadratic equations
- Roots and radicals
- Complex numbers
- Linear inequalities
- Problem-solving applications

Arnold Community College (Half Semester)

- System of linear equations
- Functions
- Quadratic equations
- Roots and radicals
- Complex numbers
- Problem-solving applications

Blair Community College (Full Semester)

- System of linear equations
- Functions

- Roots and radicals
- Quadratic equations
- Complex numbers
- Linear inequalities
- Absolute value equations
- Problem-solving applications

Walsh Community College (Half Semester)

- Functions
- Operations with roots and radicals
- Quadratic equations
- Operations with complex numbers
- Problems solving applications
- Linear inequalities

Appendix F: Sample Outline for Introduction to Statistics with the Corequisite from Arnold Community College

Corequisite Topic	College-Level Topic
Rounding numbers and percentages	Statistical terminology and designs
Measures of central tendency: mean, median, and mode	Histograms, dot plots, and box and whisker plots
Summative notation	Stem and leaf plots
	Line and bar graphs
Basic probability and calculator syntax	Probability: independent and dependent
Linear equations and inequalities	Contingency tables, discrete probability, and basic continuous density functions
Calculator syntax	Normal distributions and standard deviations
Basic notation, evaluating expressions, and roots and radicals	Confidence intervals
Equations with roots and radicals and proportions	Various forms of hypothesis testing
Basic notation	Chi-square tests
Linear equations	Linear regressions

*The corequisite classes include review of the college-level material as well.

Appendix G: Sample Outline for Quantitative Reasoning with the Corequisite from Walsh Community College

Corequisite Topic	College-Level Topic
Computing percentages and ratios	Interpreting pie charts and various graphs
Linear equations and linear equations with fractions	Personal finance and debt-to-income ratios
Computing mean, median, and mode	Histograms, dot plots, and box plots
Review	The basics of standard deviation, normal curves, and skewness
Calculator syntax and basic probability	Probability with independent and dependent events
Linear equations, slope, and equation of the line	Linear modeling
Additional practice and review	Simple and compound interest
Additional practice and review	Using logarithms to solve compound interest problems
Practice with Excel	Exponential modeling

Appendix H: Lead Questions for Student Participants

1. How did you feel about having to take math when you started community college?

2. Please tell me about your experiences in math when you were in elementary or middle school. What were the positives, the negatives, and everything in between?

3. As you progressed through math in elementary and middle school, what became more difficulty?

4. Please tell me about your experience in math in high school.

5. How did you feel starting community college?

6. How did community college start for you? Tell me about starting your first math class.

7. What do you feel led to the non-completion of your first developmental math class? Please give as many examples as possible (only for selected participants).

8. How did you do in your other math classes? (Only for those who failed their first developmental math class.)

9. What do you feel led to your success in your first developmental math class. Please be as specific as possible.

10. Please tell me about your experience with supplemental instruction (only for selected participants).

11. What were your academic career plans after passing your first developmental math class?

12. Please tell me about some of the problems you faced in intermediate algebra. Please give as many examples as possible (only for selected participants).

13. Please tell me about what led to your success in intermediate algebra. Please give as many examples as possible (only for selected participants).

14. Please tell me about successful test-taking strategies.

15. Did you ever consider enrolling in an online math class?

16. Please provide a piece of advice for students attempting to complete the college-level course that you successfully completed.

17. Please tell me about your experiences after passing college algebra (only for selected participants).

Appendix I: Lead Questions for the Lester Community College Faculty

1. A student attempts college algebra with developmental math background. What are some common problems you see? More specifically, where do you find students struggle?

2. A student successfully completes developmental math and college algebra (with a "B" or higher). The student then decides to pursue a degree in math education, which requires the successful completion of calculus and additional abstract math courses. What advice do you have for this student?

3. If a student is relying on their background in developmental math along with college algebra, what problems do foresee them encountering in courses such as pre-calculus, calculus, or beyond.

4. Let us say a student specifically lacks an understanding of geometry, a topic not covered in developmental math, how can the student prepare for calculus and beyond?

References

Krathwohl, D. R. (2009). *Methods of education and social science research* (3rd ed.). Long Grove, IL: Waveland Press, Inc.

Merriam, S. B. (2002a). Assessing and evaluating qualitative research. In S. B. Merriam & Associates (Eds.), *Qualitative research in practice: Examples for discussion and analysis* (pp. 18–33). Jossey-Bass.

Merriam, S. B. (2002b). Introduction to qualitative research. In S. B. Merriam & Associates (Eds.), *Qualitative research in practice: Examples for discussion and analysis* (pp. 3–17). Jossey-Bass.

Patton, M. Q. (2002). *Qualitative research & evaluation methods* (3rd ed.). Sage Publications, Inc.

Ridenour, C., & Newman, I. (2008). *Mixed methods research: Exploring the interactive continuum*. Carbondale: Southern Illinois University Press.

Appendix J: Samples of Teaching Practices

Example of the "Where is Everything Quiz."

1. What is your professor's name?
2. Refer to the syllabus and provide the date of the first exam.
3. Where are the online notes located? Be specific
4. Where are the written homework sheets located?
5. When are the unit 1 online assignments due?
6. What are your professor's office hours?

Examples of Team Jeopardy using Probability

- The answer is $P(A) + P(B) - P(A \cdot B)$
- Sample Question from a Team: John rolls a single fair die and flips a fair coin. What is the probability that the die lands on 4 or the coin lands on heads?
- Answer: A scenario that utilizes dependent probability.
- Sample Question from a Team: A bag contains five blue marbles, six red marbles, three green marbles, and two black marbles. We choose a marble, record the color, and do not replace it. Then, we choose another marble. What is the probability of choosing a red marble and then a black marble?

Index